Contents

Acknowledgements

The idea for a book about urban wildlife came about after discussions with Alexandra Artley. I am grateful for the work which Margaret Crowther did on the book throughout its gestation period, and to Beth Young at Blandford Press for ensuring that the material was produced in book form. The illustrations have been drawn by John Heritage: to him I offer my thanks. Mrs Lin Reynolds has typed the manuscript with her usual expertise.

Inevitably in a book of this nature, help and advice has been sought, and freely given by a number of people. I am grateful to Tim Jones (the Chairman at the time, and now a member, of Daventry Natural History Society), who was kind enough to read the manuscript and offer valuable comments. E. Rex Mathewson also read the text at short notice, offering a number of suggestions for improvement.

W. G. Teagle gave me ideas for items to be included in the 'boxes', and his intimate knowledge of the urban environment has been extremely helpful in this sphere. However, the final words are my own, and I accept all responsibility for them.

Finally, I apologise to Joan, Helen and Sarah for having 'hidden away' during difficult moments!

THE URBAN DWELLER'S
WILDLIFE COMPANION

A dunnock (or hedge sparrow) feeding a young cuckoo which has taken over its nest.

Previous page: Canada geese. Opposite: Mandarin ducks.

THE URBAN DWELLER'S
WILDLIFE COMPANION

RON WILSON

ILLUSTRATED BY JOHN HERITAGE

BLANDFORD PRESS
POOLE · DORSET

First published in the U.K. 1983 by Blandford Press,
Link House, West Street, Poole, Dorset, BH15 1LL.

Copyright © 1983 Ron Wilson

Distributed in the United States by
Sterling Publishing Co., Inc.,
2 Park Avenue, New York, N.Y. 10016.

British Library Cataloguing in Publication Data

Wilson, Ron
    The urban dweller's wildlife companion.
    1. Urban fauna—Great Britain
    I. Title
    591.9'41   QH541.5

ISBN 0 7137 1324 0

Typeset by August Filmsetting, Warrington, Cheshire

Printed by Butler and Tanner Ltd., Frome & London

Introduction

An area of offices, shops, factories, houses and other miscellaneous buildings is, to most people, their view of the urban environment. Many who live in the country see it as a sterile wilderness; an area which presents no niches for wildlife; an area devoid of flowers, mammals, birds and butterflies.

Yet, although that might seem an appropriate view of the inner city areas, it is generally a distorted one. Even in the concrete heart of our most densely populated areas there is wilderness, and wilderness supports wildlife. It may not match the diversity in certain country areas (even that is not as secure as it was thirty years ago), but there is a range which, when the urban dweller becomes aware of it, will not only surprise but also delight him.

Indeed, because of the nature of the environment, wildlife has become adaptable. Walls have been colonised, spaces between paving slabs taken over, ledges commandeered and dustbins scavenged. So wide and so surprising is the range of wildlife which occurs in towns that this book has only touched the tip of the iceberg.

Nevertheless we do hope that the book will encourage the non-naturalist in particular to explore his 'barren wastes'. Such exploration has already led to the discovery of a diversity of shape and form which most people thought the urban situation incapable of supporting.

Ron Wilson
Weedon 1983

FORSAKEN LAND

Waste ground and rubbish tips

The lucky country dweller may feel sorry for the person subjected to a life in the urban environment. Country people often see the town as an almost impenetrable maze of brick and concrete, with every spare metre of land taken up with offices, factories and shops. These assumptions are not true of most built-up areas, however. Apart from the parks and other planned open spaces, there are many areas of 'forsaken land'. These include a variety of derelict areas, building sites, car parks, neglected land around factories and other buildings, waste land near rivers and docks and, although they are not so numerous nowadays, bomb-sites. Although some of these might seem insignificant or inhospitable, collectively they represent a valuable resource in the built-up area. They present a veritable oasis for plants and animals.

Ragwort provides the cinnabar moth with a suitable supply of food. Cow parsley, a larger umbellifer, presents a delightful expanse of white flower heads, proving a useful and much sought after platform for many insects. The coal tit comes to feed. Timothy grass flourishes in many habitats.

Plant life

The destructive power of fire in history is well-known, and one of the great city of London's most disastrous events was the Great Fire of London, in 1666. Throughout the centuries all built-up areas have suffered fire to a greater or lesser extent. Many urban areas were badly affected during the Second World War.

After both the Great Fire and the War new species of plants have taken up residence. Once such species become established they spread, and colonise new areas. For example, after the Great Fire, Canadian fleabane (*Erigeron canadensis*) was first recorded in the city. This plant has been uprooted from its natural habitat of waste areas and is now also found in many gardens.

A great opportunist

The great opportunist of all time must surely be the rosebay willow-herb (*Epilobium angustifolium*). It was one of the first species to flower on the bombed sites of London after the Second World War, adding a much needed splash of colour to the dereliction all around. It is hardly surprising that at the time the citizens of London dubbed it the 'bomb weed'. Now it rejoices in the alternative name of fireweed. This is equally appropriate, for it is one of those species capable of colonising ground smitten by fire.

A tall, erect plant, which may reach a height of up to 1.2m (4ft), it is now found in many habitats apart from the waste areas which it initially colonised. The four-petalled rose coloured flowers are arranged towards the end of the flower spike. Its ability initially to colonise

and then maintain and increase its occupied territory is due, in the first instance, to its prolific seed production, and then to its underground stem. It has been established that a single plant can produce as many as 85,000 seeds. Each is light, almost weightless, with a tiny parachute-like structure attached to it. Such a mechanism allows the wind to carry it widely to new areas. Once established the underground rooting system ensures that the plant will spread over the maximum amount of territory.

Although most people have derided this persistent weed, the young shoots can be cooked and served like asparagus. Leaves can be picked, prepared and used as a vegetable. Where large numbers grow, and bees collect the nectar, a good honey will result. How well this plant has done, too. When it was noted by botanists a hundred years or so ago, it was considered uncommon. In those hundred years its ability to colonise such a wide habitat has changed its classification to common. Many reasons for its sudden increase have been given, including the movement of building materials from one area to another. It has also been suggested that the plant has undergone some genetic change, which has made it more versatile.

A plant from Europe
The Oxford ragwort (*Senecio squalidus*) has taken advantage of the continually disturbed soils of new building sites and disturbed areas. As long as the soil has a good supply of lime, which building rubble will usually provide, the plant will soon become established. Walls rich in lime have provided a niche for the Oxford ragwort. Reaching a height of between 15 and 30cm (6 and 12in) it has flower heads which are usually no more than 20mm ($\frac{3}{4}$in) across.

Seeds with 'parachutes'
Although there have been some species which are either new to the British list of plants, or have dramatically increased their range, there are other common species which have always grown on waste ground. Once the dandelion (*Taraxacum officinale*) manages to become established, it is not easy to get rid of. Initial colonisation by dandelion plants will generally be the result of wind-blown seeds. The well-known dandelion clock consists of many of these light, fluffy seeds. Each seed has its own efficient parachute, which carries it away from the parent plant. Once it lands on the soil, the seed, being heavier than its parachute, becomes slightly embedded, and, provided conditions are right, it will start to germinate, and a new dandelion plant will eventually grow.

Beneath the ground the dandelion develops a thick tap root. It is from this that the leaves grow. The well-known yellow composite flowers are borne on long,

hollow stems, inside which there is a milky coloured liquid. Although the main flowering season is from March to October, dandelions will be found flowering in many places even in the most severe weather.

The plant may get its name from the French *dent-de-lion*, a reference to the indentations of the leaf, which bear a rough resemblance to teeth – if you use your imagination to the full!

The early flowering coltsfoot
Where there is disturbed ground one of the earliest species to grow is the coltsfoot (*Tussilago farfara*), as long as the soil retains some of its moisture. It has been dubbed coltsfoot from the shape of its leaves, which incidentally do not appear until after the flowers have faded. It is particularly conspicuous in late winter and early spring, because of its yellow flowers which have the advantage of exhibiting themselves without the problem of being overshadowed by leaves. The flowers are not unlike large daisies. If the weather is cold, they will usually close, a feature which is peculiar to some coltsfoot plants at night.

Coltsfoot was much used in earlier times, and was relished because of the medicinal properties it possessed: it was useful for curing colds and other complaints associated with them. Its Latin name of *Tussilago farfara* gives a clue to its curing powers. *Tussis* means a cough, a reference to its importance as a cure for chestiness. Today, as herbal remedies are regaining some of their popularity, coltsfoot is still used in many

1 DANDELION
The innumerable light seeds of the dandelion ensure that dispersal will be effectively carried out. Young dandelion leaves are collected for salads, and dandelion coffee is now being drunk by more and more people.

2 STINGING NETTLES
Much despised, the stinging nettle is important to many animals. The small tortoiseshell, red admiral and peacock butterflies lay their eggs on the plant's leaves. When the caterpillars emerge they have a plentiful supply of food.

4 CREEPING BUTTERCUP
Although the buttercup does produce seeds, it effectively increases by creeping underground stems, which send up leaves at regular intervals.

3 ROSEBAY WILLOW-HERB
Its alternative name of fireweed accurately describes the ability of the rosebay willow-herb to colonise areas which have been ravaged by fire.

5 CUCKOO PINT (WILD ARUM)
The wild arum has many common local names, including cuckoo pint, jack in the pulpit and lords and ladies. The smell which the plant gives off attracts small flies which bring about pollination.

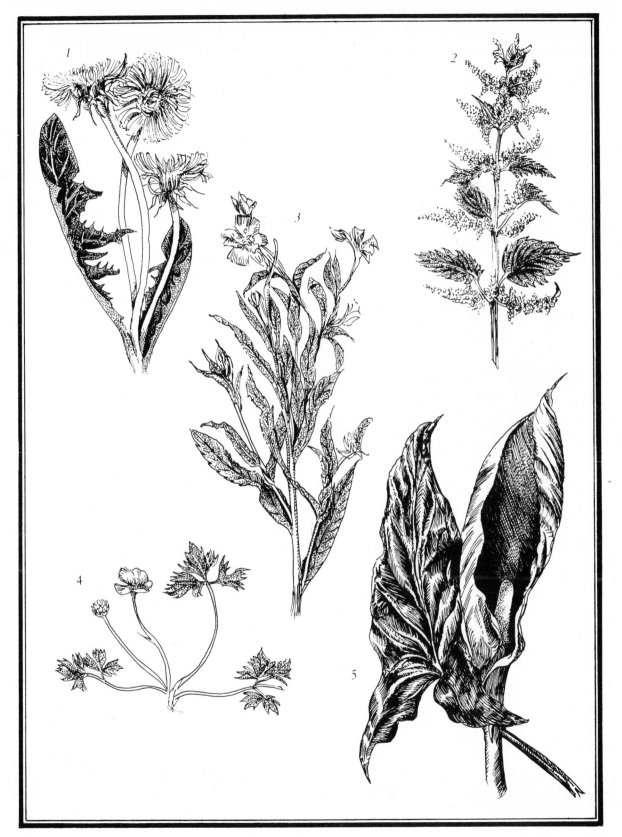

ways, but especially for the relief of respiratory diseases. For the preparation of a herbal wine, either leaves or flowers can be used. Coltsfoot is also manufactured into sticks and sold in many health shops. The leaves were used as a tobacco substitute and are still the basis of a herbal smoking mixture.

Inconspicuous flowers

Waste ground will probably support many inconspicuous flowering species, some of which will also be found in other areas. Where large numbers of fat hen (*Chenopodium album*) occur they will often go unnoticed because of their drab, tiny green-coloured flowers. They will, however, produce large numbers of seeds which will eventually give rise to even more plants. So numerous are the seeds that it has been estimated that a single plant may produce as many as 100,000. Growing to a height of about 100cm (39in),

> **TEASEL** (*Dipsacus fullonum*)
> At one time the teasel was grown because the seed heads of one particular species – Fuller's teasel – were used for napping cloth. The word teasel comes from the Old English *taesan*, which means 'to tease'.
> The Romans called the plant *Lavacrum veneris* which meant 'basin of Venus', a reference to the fact that where the leaves join around the stem they form a small 'basin' which collects water. Insects which drown here dissolve, and their remains are taken in by the plant as part of its food.
> A biennial, the plant grows in suitable places along railway lines and in waste places. During its first year it grows a rosette of leaves, which closely hug the ground. The tall stems are not produced until the second year.

the plant is also able to survive and produce seeds when only 10cm (4in). As with the stature of the plant, the seeds are also extremely variable, even on the same specimen. They may range from black to pale yellow. There seems to be a correlation between seed colour and germination. Darker seeds tend to take longer to grow than lighter ones.

Considered a persistent weed, because of the viability of its seeds, it has been growing with and around man since the time of the Stone Age. Although of no commercial value now, the seeds were once used as a substitute for corn, but how nutritious they were we have no idea.

Common plants

A plant equally at home on the rubbish tip and on cultivated land is the groundsel (*Senecio vulgaris*). An annual species common throughout Britain, it grows to a height of between 15 and 30cm (6 and 12in). After being fertilised the yellow flowers will fade, and in their place will be the familiar fluffy seed heads, so typical of this and other members of the daisy family, which includes common species like dandelion and coltsfoot. Because of the light seeds and the added buoyancy which the fine hairs (*pappus*) give them, the plant has been carried far and wide.

In spite of the creeping thistle's (*Cirsium arvense*) attractive down, which looks harmless as it floats serenely on the summer breeze, when the plant becomes established in gardens it is difficult to eradicate and becomes a nuisance. Any small part of the root which becomes detached from the main root will start to grow and form a plant in its own right.

In bloom in summer, the smallish purple flower heads are borne on the stem in loose clusters. These seeds too are supplied with a long pappus hair, which gives them a parachute effect when they are carried by the wind.

Early fungi

An edible fungus, the shaggy ink-cap or lawyer's wig (*Coprinus comatus*) seems to flourish on rubbish tips. It is one of those species which can be found earlier in the year, and the first ones may be discovered as early as May.

The cap of this particularly striking toadstool, unlike that of the common mushroom, is longer than it is broad. Although there is great variation in height, it may reach 30cm (12in). As it ripens it changes colour from an almost unnoticeable pink through to a striking black, then it liquifies, and the drops of liquid which fall away contain the spores which, if they grow, will perpetuate the species.

In need of support

Where supports are available on waste ground, many climbing species will colonise the area. One which is frequently found is the woody nightshade (*Solanum dulcamara*). Although it will grow upwards provided that it has something against which to lean, it does not have any adaptations for this habit: there are no hooks or tendrils, which many other climbing species exhibit. It will trail over the ground as well as grow upwards. A well established species, many plants reach a height of 1.2 to 1.5m (4 to 5ft).

There is a variety of form exhibited by the leaves. Those nearest the ground are usually heart-like in outline: others are more lance-shaped. All leaves have an un-serrated edge and a smooth surface. Although the flowers are purple in colour, they are not particularly large. The most conspicuous part of the plant is the berry which develops once the flowers have been fertilised. At first the fruits are green, but gradually they go through a series of colour changes, from yellow to orange, finally taking on a bright scarlet colour. Its alternative name of bittersweet is from its Latin name *Solanum dulcamara* – *dulcis* means sweet and *amarus* bitter. Apparently the berries taste sweet at first, but then bitter. It is not poisonous, and has its uses in herbal medicine, having been used in the relief of rheumatism. It is also related to the potato.

ELEPHANT HAWK MOTH

Because of its large nature the caterpillar of the elephant hawk moth is usually seen more often than the adult, and is sometimes mistaken for a small snake. Where willow herbs, bedstraw and cultivated fuchsias are found, it may be discovered feeding towards the end of summer. The adult makes its appearance sometime in June, and may be encountered at dusk, when it visits flowers.

Trampled species

Although many waste areas are allowed to grow wild, parts are often trampled over, and the plants which have their leaves arranged in a rosette formation are particularly well adapted to such an existence. Those species which spring to mind include the daisy (*Bellis perennis*) and the plantains (*Plantago* spp.), plants which do very well on the lawn as well, where similar difficult conditions occur. The seeds of such species are not light, and so rely on a different method of dispersal from those described earlier.

Of the three common species of plantain – greater (*Plantago major*), ribwort (*P. lanceolata*) and hoary (*P. media*) – all are likely to be encountered on waste ground. Such species are typical of many habitats. The shortest of the three is the greater plantain. This reaches a height of between 10 and 30cm (4 and 12in). Like the daisy it will flourish where areas are mown or trampled, and it will establish itself on paths.

Flowering between summer and autumn, the flowers on the plantain are extremely dingy and would not be taken for flowers at all by most people. The leafless stem which bears the flowers arises from the rosette of egg-shaped leaves. Although the flowers become more conspicuous when the brown-coloured stamens are present, they are still not particularly evident.

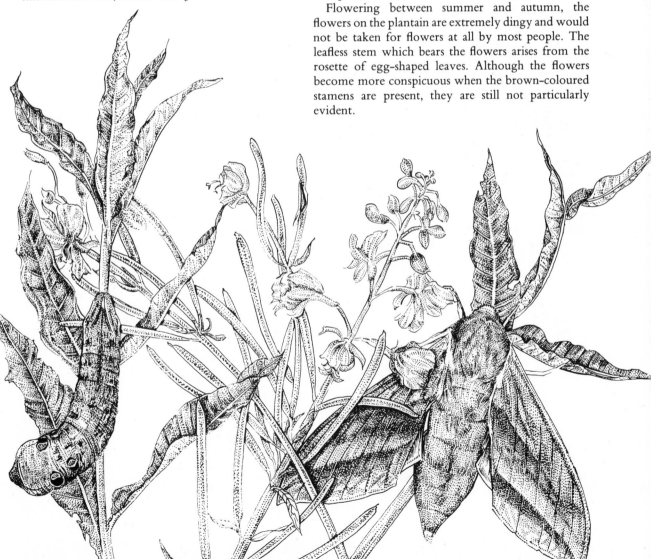

The seeds which appear later in the year are black and hard. They will be scattered by various animals, including birds. The seed-laden stems attract goldfinches, and they may spend a great deal of time swinging on them, scattering the seeds in all directions.

Ribwort plantain is also called 'cocks and hens' and 'cocksheads', because of the arrangement of the seed heads. Growing to a height of up to 45cm (18in), the species is the most common of the three plantains. Although its leaves resemble those of the other species, they are much narrower than those of the greater plantain. Slender and lance-shaped, the leaves have between three and five prominent veins on the under surface. Particularly resistant to mowing, this species of plantain will also be found in waste areas, including places where the grass grows tall.

The leafless flower stalk bears club-like structures, usually between 2.5 and 5cm (1 and 2in) long. Each structure is made up of a number of flowers, which are readily visible because the flower stalk pushes its flower heads upwards. Once the fruits have developed the seeds – brown and hard – have a rough texture.

The third of the three species, hoary plantain, has leaves formed into a close compact rosette, hugging the ground and effectively preventing other plants from growing underneath. Because they are covered with very soft, white hairs, the leaves have a silky feel to them.

Composite flowers

Several species of Compositae, which resemble each other, are common species on waste ground. These are stinking mayweed (*Anthemis cotula*), scentless mayweed (*Tripleurospermum maritimum*) and wild chamomile (*Matricaria recutita*). All three have daisy-like flowers. Scentless mayweed has the largest flower, which may be as much as 28mm (1.5in) across.

Of the three species, only wild chamomile has flowers which give off a pleasant aromatic scent. The feathery leaves produce a similar smell when they are rubbed between the fingers. As is to be expected, this is not the case with the stinking mayweed. As its name implies, the leaves give off a rather unpleasant aroma when crushed, and the plant was particularly disliked by our ancestors, where they had to hand-weed crops. Its disagreeable odour, however, was not the only thing which displeased our forefathers. According to some writers, when it was being pulled from the fields the plant left their hands and the bare parts of their bodies blistered. Handling large numbers of the weeds, and being attacked in this way, gave rise to a great deal of pain, with the result that men were unable to work for many days.

All three plants can be found flowering throughout most of the summer. Both mayweeds will often be found in bloom well into the autumn if the weather remains favourable.

Wild chamomile, unlike stinking mayweed, has been a very useful plant. Indeed, it is still used by herbalists, providing in particular the basis for a very pleasant tea. It was also used for making soothing poultices for wounds.

Dispersal of seeds

There are many important times in the life history of every plant. Without fertilisation no seeds would be produced; but it would be no good having seeds if they could not be dispersed in an efficient way. Some plants, like the dandelion, rely on the wind to carry their seeds to pastures new. Others rely on birds and mammals. Two of these are cleavers (*Galium aparine*) and lesser burdock (*Arctium minus*).

A common species, cleavers is found in many habitats including waste ground. It has alternative names of goose-grass and sweethearts. It clings to other plants by means of the mass of tiny hooks with which it is covered, which are found not only on the leaves and stems, but on the fruits when these appear.

The leaves grow in whorls around the stem, the number in each whorl varying from as few as four to as many as ten. An annual plant, it makes good progress within a season. The square stems sprawl and trail for a couple of metres (6ft) or more.

The flowers, very small and almost inconspicuously greenish-white in colour, appear in June or July. Once fertilised, the fruits will start to develop. At first they are green, but as they ripen they take on a purplish tinge. Once they are ready, their surface is covered with a mass of small, but distinct, hooked bristles, which help them stick to clothes, feathers, fur and so on. In this way they can be carried some distance from the parent plant.

Burdock also spreads its seeds by means of tiny hooks. Two species of burdock may be found growing in waste places. These are the lesser burdock and the wood burdock. Of these two, the lesser burdock is extremely common in waste places. The latter also occurs in dry open woods and along roadside verges.

A biennial, the burdock's growth starts in one year and will be completed with the production of flowers and seeds in the second year. Generally the plant averages between 0.9 and 1.2m (3 and 4ft), but in secluded spots, especially where its growth is not interrupted by other species, it may top 2m (6ft). The branching stem is extremely stout. Close to the ground the leaves have a heart-shaped outline. Underneath they have a cotton-like texture, and the edges are irregular and wrinkled. Towards the top of the plant the leaves are smaller.

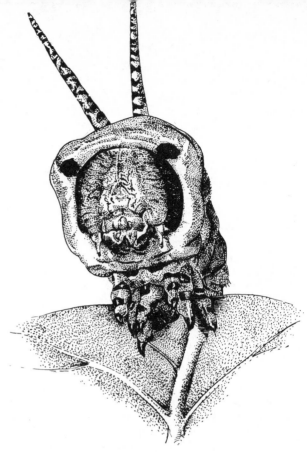

PUSS MOTH

The nocturnal puss moth is on the wing from May to June, when it may commonly be seen in many haunts, including parks. The white transparent wings have a variety of grey markings. The caterpillar hatches from eggs which the female laid on poplar and willow, and sometimes aspen. During the early days of life, it is dark in colour, generally black. As it moults it becomes brighter.

The plant will be in flower from June to September. When these flower heads ripen, and give rise to the seeds, there are numerous sharp hooks on each of the bracts. Virtually anything which comes into contact with them will carry them away. Once in a suitable place, they may begin to germinate.

Food for animals

As Oxford ragwort and rosebay willow-herb spread to new areas, they were quickly followed by the animals which feed on them. One of those which took advantage of the increased food supply was the elephant hawk moth (*Deilephila elpenor*). Although the night-stroller may come across this large, beautiful insect, it is more likely that it is the caterpillar which will be encountered. Once it has been identified, it is highly unlikely that the larva, which feeds on the leaves of rosebay willow-herb, will be mistaken again. Large and conspicuous, measuring up to 8cm (3in) in length, the larva's main body colour is variable: in some instances it is green with a combination of black dots, in others black and brown. The rear end of the body is surmounted by a horn-like appendage.

Once the larva has pupated, the large and distinctively marked elephant hawk moth will emerge. With a body length of around 6.5cm (2.5in) the insect is marked with pink and green. A walk around waste ground at dusk in June will generally reveal the female flitting around willow-herb plants, seeking out a suitable one on which to lay her eggs. If rosebay willow-herb is available it is not interested in any other plants. If it cannot find this particular species, another willow-herb will do.

Names from food plants

Other hawk moths, and in particular their caterpillars, may find a source of food on waste ground plants. Provided that the right plants are available, if the females are around, they will lay their eggs. Many hawk moths have derived their name from their main, although not necessarily sole, food plant – species like the convolvulus, poplar, privet and lime.

Most species of moths are nocturnal, coming out at night to feed, protected to some extent by the darkness. Yet, in the daytime, because they are at rest, they need to be protected from the would-be ravages of predators. Camouflage is therefore a very important aspect of their colouration. Puss moths (*Cerura vinula*), for example, will spend the daylight hours resting on tree trunks and fences.

Wild and cultivated cabbages

Man's crops have provided the large (cabbage) white butterfly (*Pieris brassicae*) with an abundant supply of food. It has taken advantage of this and is now considered a pest. In the wild, however, and where cabbage plants – its chief cultivated crop – do not grow, it will look for other plants. The young caterpillars need members of the cabbage family, and these include the hedge mustard (*Sisymbrium officinale*). Where waste ground occurs close to allotments and gardens, escaped nasturtium and radish might be found and provide food.

The caterpillars pupate in late summer, overwintering in the 'safe' state of suspended animation, as a chrysalis. At what time they will emerge as adults in the following spring will depend on the weather. Generally the first adults are to be seen on the wing in the early days of May, having emerged from the pupa.

Once the butterflies have mated, the female will seek out the necessary food plants. Sometimes the eggs are laid on the upper surface of the leaves; sometimes they are underneath. The eggs appear to be positioned in a haphazard bunch, as if the female was in a hurry to complete the operation. After emerging from the eggs the caterpillars stay together. At first they only attack the surface of the leaf, scraping away the greenery. After they have shed their first skin, they get to work with a will, and will eat holes in the leaves. It is not unusual to find them in a kind of formation. Situated around the edge they devour the leaf as they move towards the centre. Having grown to about two-thirds of their full size, they move away from their fellows, and feed on their own.

When the first brood is ready to pupate, they can be found generally attached to either the stems or leaves of the plant on which they feed. Others will leave, and pupate on fences and under window ledges. The percentage which pupate varies from brood to brood, and from year to year.

The offspring of this brood will emerge as adults in August and can be seen flying usually until the middle of October, sometimes later if the weather is mild. Before this time they will have mated, and the female will have laid her eggs. Although the adults will continue flying well into autumn, they will gradually die as the cold comes.

A parasite

A threat to the caterpillars before pupation is the female ichneumon fly (*Apanteles glomeratus*) which lays her eggs inside their bodies by inserting her hollow 'ovipositor' through the skin. The eggs are then released through this. As each ichneumon larva develops it gradually eats the inside of the caterpillar, so that death occurs before pupation is reached. When the ichneumon fly lays her eggs inside the caterpillar's body, she may deposit as many as one hundred eggs. The ichneumon fly larva does not damage those organs in the caterpillar which keep it alive. The ichneumon grubs reach their fully grown state at about the same time as the caterpillar is ready to pupate. By then the ichneumon fly larvae have eaten all the inside of the caterpillar, and they leave nothing more than a skin. At this stage they are ready to pupate, and leaving the caterpillar they spin their own small yellow cocoons around themselves.

Of course not every caterpillar is attacked by these ichneumon flies. In spite of this, however, they have other enemies. A chalcid, an extremely small insect, lays its eggs not in the caterpillar but in the chrysalid. As the grubs feed on this they eventually kill it.

Widespread cuckoo spit

It is more than likely that you have been confronted with the familiar blobs of froth on the stems of a wide variety of plants. You have perhaps wondered what this collection of whitish frothy material, not unlike human spittle, conceals. Wonder no more: for you can look beneath the protective froth to find the nymph of the froghopper bug (*Philaenus spumarius* and *Neophilaenus lineatus*). From late spring through the summer months it will be noticeable on a wide variety of herbaceous plants. People are often confused by the name 'cuckoo spit', and ask the inevitable question, 'Why?'. The question can be answered quite simply: the froth usually appears around the time that familiar summer visitor, the cuckoo, arrives.

The female froghopper bug lays her eggs on the plant in autumn, and they remain in this inactive state until the following spring. Once the young have hatched, they will make their way up the freshly growing plant, and bite into the stem. Once their biting mouthparts have penetrated the stem they are able to take their fill of watery sap. In these young and succulent plant stems the froghopper nymphs should find an abundant supply of food.

In such an immobile position they would probably be quite vulnerable to their predators; but the water in the sap passes through the body, and out from the anus. When air is added to the liquid froth is produced. It serves two purposes. First, it offers the otherwise susceptible nymph a very necessary form of protection against the attacks from would-be predators. Second, if it was not covered by the froth the nymph would dry up quickly in direct sunlight.

Once the nymph has found a suitable plant on which to feed, it generally stays in one position. Sometimes, perhaps because of a lack of sufficient food, or for other unknown reasons, it has to move. Its first activity, once established at a new feeding station, is to cover its body with the well-known froth.

These froghopper bug nymphs will feed from spring to around mid-summer, and once the nymphs have taken their fill, and are fully grown, they will be ready to make the transition from juvenile to adult. Once the adult has appeared, as its name suggests, it is capable of movement. Called a froghopper, it looks like and tends to make movements not unlike the better-known amphibians. Many adults, however, also remain on one plant for a long time, taking the necessary nourishment by inserting a proboscis into the stem to remove the liquid.

The most common species to be found in most areas is *Philaenus spumarius*, and although it does occur frequently on rosebay willow-herb, especially in neglected areas, it has been found feeding on no fewer than four hundred different plants. It would seem that the bug is quite harmless, because the foam the nymph produces does no damage to the plant. However, where they occur in large numbers on plants, by taking vital nutrients they will deprive the plant of food needed for growth. Thus some plants might not grow to their full capacity.

Birds seek food and homes
Undoubtedly as long as there is food, and the greater the variety the better, birds will come to waste areas.

They may come in to feed, or if suitable conditions are available they may nest. At certain times of the year weeds will provide the seed-eaters with a good supply of food. If there is also small animal life, the insect-eaters will find nourishment for themselves, and in the breeding season for their hungry nestlings.

Common urban dwelling birds, like house sparrows (*Passer domesticus*), blackbirds (*Turdus merula*), starlings (*Sturnus vulgaris*) and blue tits (*Parus caeruleus*), will usually find a continuing supply of food. Given a place to nest, the sparrow will build the home for its eggs and youngsters. Being opportunists, and undoubtedly the most common urban bird, house sparrows will make use of any available nesting site. In rural situations they generally confine their sites to trees and bushes, but in built-up areas they may come into buildings if the need arises. This is especially true of derelict areas, where buildings remain empty and the nest site undisturbed.

Unlike some of the other birds which frequent the urban environment, male and female house sparrows have distinctly different markings, and so are easy to distinguish. The hen bird has dark brown plumage above, and this is broken up by a series of brown streaks. Below she is greyish. Although the male's general colour is similar, he is enhanced by the addition of a black area on the throat, and his head is embellished with a grey cap.

Prior to mating there is a courtship ritual which

BLACKBIRD
Highly adaptable, the blackbird seems to be able to find a niche in almost every situation, and because of this ability the population has increased dramatically during this century. The bird is extremely common, and it and the chaffinch vie for the title of Britain's most common species.

STARLING

Few birds are as unwelcome as the starling, and yet careful examination will reveal a particularly attractive feather pattern. Able to live by 'scavenging', the bird has managed to thrive and expand its activities in the urban environment. Flocks of starlings descend in extremely large numbers onto their roosting sites at night.

ornithologists have termed the 'sparrow party'. Once a cock has his eye on a female, he will start to serenade her, and then make definite bowing movements. As he becomes more daring he moves towards the hen. It is not unusual for her to repel his advances by pecking at him, apparently uninterested in his attentions. Whether simply inquisitive or hopeful of also finding a mate is a matter for conjecture, but then other cock birds join in. The female accepts their 'goings-on' for a while, but will soon tire of it and fly off; but she has not escaped from them, and they will take off and fly after her.

Once mates have been selected, both birds will take part in the nest-building operation. The nest is not a very tidy affair, consisting of a mixture of dried herbs, most of which is usually grass. Once completed, egg-laying takes place. The season covers the period from April to August, with the female producing between three and five eggs in a clutch. There may be as many as three broods in any one season.

With the population explosion in urban situations, the sparrow has taken advantage of every conceivable opportunity to obtain the food necessary for survival. It has learnt, although rather ineptly, to remove the tops from milk bottles, to peck at nuts, and to attempt to copy the woodpecker by searching trees for insect larvae, even though it is not at home here, as it is when hopping along the ground in search of a supply of food.

Hide and seek

A wide variety of rubbish tends to collect in waste areas. Pieces of tin, corrugated iron and asbestos provide a secure and sheltered, although sometimes temporary, home for many species of animal life, including small mammals. Regular visits to such mini-habitats will reveal that a whole range of creatures has found a place in which to shelter, and if they are lucky and not disturbed too often a secluded breeding site.

Seasonally the corrugated tin and asbestos habitat will change. Winter occupants will include small mammals, such as voles; in summer it will be home to lizards and many insects, and perhaps to slow worms as well in some places.

Whereas most animals do very little, once a vole has moved in it does its best to make its new found home habitable. Numerous tunnels or runs appear, and if it stays to breed beautiful nests will also be constructed.

Food will be collected and droppings accumulate, both signs that the area is occupied. Other small mammals – shrews and mice – may find the asbestos and tin sheets to their liking. If you do discover these materials, and want to investigate the population beneath, do it cautiously, and do not forget to replace them carefully.

The rubbish tip

Apart from waste land, in any town the 'uninteresting' rubbish tip, especially where household waste is dumped, may be a veritable treasure house providing a great deal of valuable food for a wide variety of animals, birds and insects. Among these are some perhaps unexpected inhabitants – the slow worms, grass snakes and crickets; as well as those which we would expect to find – the rats, starlings and cockroaches.

Birds round the tip

By far the most conspicuous of the rubbish tip animals are the birds. Although at one time we would only have expected to encounter gulls around the coast, they are now found in many towns far from the sea. The majority of these will find a plentiful supply of food material on the rubbish dump, for they are true scavengers.

Although gulls seem well established in the urban situation, they were not the original scavengers; this honour went to the kites and ravens, especially in London. Gulls were not noted in any numbers in the capital until towards the end of the nineteenth century. It was the advent of adverse weather conditions at this time which drove them inland in their desperation for an alternative source of food. Now it is during the winter months that the number of gulls living and feeding in towns increases dramatically.

Black-headed gulls (*Larus ridibundus*) do breed in towns, and they have been nesting in London since 1946. Sewage farms, rather than rubbish tips, are their preferred nesting sites. The bird also has the distinction of being the first gull to make the town, and in most areas it is by far the most common.

Given suitable conditions, the herring gull (*Larus argentatus*) will also nest, although it takes to ledges and roofs, where its droppings may cause a nuisance to the

LESSER BLACK BACKED GULL
For many years the lesser black backed gull came inland to built-up areas in winter in its search for food, but it did not start to breed here for many years.

residents. Although the lesser black-backed gull (*Larus fuscus*) manages to breed in towns, the great black-backed gull has not yet achieved this distinction – although someone will probably prove me wrong! Many of the gulls favour breeding sites which are close to the sea, and this is true of the kittiwake. It is only relatively recently that it has sought out the urban environment as breeding quarters, and even so it is only in selected parts of Britain, mainly in northern England, that it has found town conditions suitable.

The ubiquitous starling (*Sturnus vulgaris*) finds a good supply of food at the rubbish tip, and like the gulls it is a scavenger. During the daytime the number of starlings in any city area appears relatively small. Indeed most will have left – at least temporarily – to

CARRION CROW
As part of its name implies this crow favours carrion in its diet. In addition to feeding on dead material, it may attack live animals from time to time. The carrion crow is a true scavenger, taking almost anything which it comes across.

seek their fortune in the surrounding countryside and on the suburban rubbish tip. As night-time arrives, the scene changes and thousands come together in city squares, congregating on ledges, particularly where older buildings are found.

As with gulls the starling problem in some cities is a relatively new one. The bird, unlike other species, has managed to increase its numbers in the last hundred years or so. W. H. Hudson noted that before taking to buildings they roosted in trees around lakes within the city. Gradually, during this century, there has been a population explosion, and any suitable roosting site has been grabbed with ever-eager claws.

The greatest number of starlings will be found in the

KESTREL
Birds of prey like the kestrel find a variety of live food in built-up areas in the form of small mammals. The pointed bill effectively deals with the struggling creatures. The kestrel's hovering flight is a valuable attribute as it looks for signs of movement in the grass below.

city during the autumn and winter. At this time their numbers are swollen by birds which come in from the country. Although winter visitors arrive from Russia, these birds do not join their urban relatives, but seem to replace those in the country which have moved to the town.

A competent mimic, the starling is able to copy the sounds made by many other species. This is probably to be expected, since the bird is related to that much more famous mimic, the mynah bird.

Partaker of carrion

Resident in towns for a very long time, the carrion crow (*Corvus corone corone*) feeds on carrion, as its name aptly suggests. It manages to make use of material which has been discarded by other animals, including man; but dead material is not the only food to feature in its diet. A large bird, measuring 45cm (18in), when the need arises it is not averse to attacking and killing domestic animals. It is no accident that most of the birds which manage to survive in the urban environment are opportunists. The carrion crow is no exception. Indeed there seems very little which it will not tackle. Although man now keeps his cities immeasurably cleaner than his ancestors by carefully disposing of his refuse, the carrion crow still scavenges for its food and still manages to find a great deal of material, despite competition from other species. Unlike other scavengers, when carrion crows frequent parks and other spaces where wildfowl occur they will attack and kill these birds. It is perhaps understandable that the authorities need to keep a careful watch on their population.

If conditions are suitable they may find an acceptable breeding site in and around rubbish tips; but if they do not find a nesting place here it is likely that they will seek out other suitable places in various parts of the town.

Once paired the cock and hen birds will set to work to construct a nest. Extremely bulky, it consists of a large number of twigs, and the structure will eventually be lined with an assortment of dead leaves, dried grass and, where the crows can get it from outside the town, sheep's wool. The eggs – three to five in number and light green or blue in colour, spotted with dark grey-brown markings – are laid in April or May. After the female has incubated them for nineteen days, the hungry nestlings will be fed by both parents.

Snake in the grass

Particularly where there is a good supply of decaying leaves or rotting material, the female grass snake (*Natrix natrix*) will lay her eggs. Having spent the winter in hibernation – perhaps somewhere in a suitable place in the refuse tip – the grass snakes will find partners in the spring. After mating the female will search out a suitable spot to lay her eggs. Each clutch consists of between twenty and thirty white eggs. Not unlike those laid by some birds, they can be distinguished from them by the fact that the outer shell is parchment-like. Not a particularly 'good' mother, the female may stay around the area where the eggs are laid, perhaps making some effort to ward off would-be attackers, but she does not incubate the eggs. Temperature is an important factor in the length of the incubatory period. When it is right the eggs may hatch within six weeks: if it is not correct it may be ten weeks before the young grass snakes emerge.

Once the young hatch out the female takes no active part in their protection. They move off in search of their own food. When they are born they are between 15 and 20cm (6 and 8in) long, and their food has to be in small portions. At first various grubs, earthworms and soft-bodied invertebrates will feature in the young snake's diet. On average an adult mature male will attain a length of some 60cm (24in): females are about half as long again, measuring about 90cm (36in). As the snakes increase in size the food changes, until they reach the stage where they are able to cope with larger prey. Although it is generally held that their main food consists of frogs, they adapt to the urban situation, taking newts and young voles.

Although the grass snake can inflict a painful bite, it does not possess poison fangs, and so is unable to paralyse its prey. It swallows its food whole, using the teeth which have a backward rake to push it down the throat. As the animal makes its way towards the grass snake's stomach, the unfortunate victim dies from suffocation.

Because the grass snake is not poisonous, it is completely harmless to humans.

> **KESTREL** (*Falco tinnunculus*)
> This bird of prey is no stranger to the town. In fact it can be seen over a large number of towns. The bird first bred in London as long ago as 1931, and there were at least ten breeding pairs in 1977.
>
> During its nesting periods, it has selected some of the most prestigious sites, especially in London. Here it has taken in The House of Lords and the Savoy Hotel, to name but two.
>
> The effects of the destruction by the war helped kestrels in their quest for new sites, particularly for feeding. With the desecration of large numbers of buildings, the open areas not only attracted small mammals seeking refuge, but also the kestrel seeking out the hiding mammals!

A legless lizard

People who see the slow worm (*Anguis fragilis*) normally classify it with the snakes, and often label it as venomous. In fact, although a reptile, it is not a snake but a legless lizard.

Provided there is food, the slow worm – or blindworm as it is sometimes called – will be found in damp or drier situations. It is likely to find a plentiful supply of food in the rubbish tip, taking insects and insect larvae as well as earthworms and slugs.

When the slow worms mate in spring, the male takes a firm hold of the female with his mouth. After mating the eggs develop inside the body of the female. She will lay these in late summer. They will be at such a state of development at this stage that the young slow worms will hatch as soon as the eggs have left the female's body. They may be as few as five or as many as twenty-five. At birth they have a distinctly silver tinge to the scales, and there is a recognisable darker line which runs along the whole length of the back.

In winter the slow worm hibernates, the position selected depending on what is available. A hole (either a suitable crack in a log or a hole in the tree) will be adequate. Where soft ground is available it may use its ability to burrow. Failing this, the slow worm will probably satisfy itself with the insulation which a pile of leaves will afford. Like the grass snake, it is completely harmless to humans.

STAG BEETLE

No one is quite sure why the male stag beetle has such large jaws. Although they appear as a fierce appendage, they are ineffective because the muscles which control them are very weak.

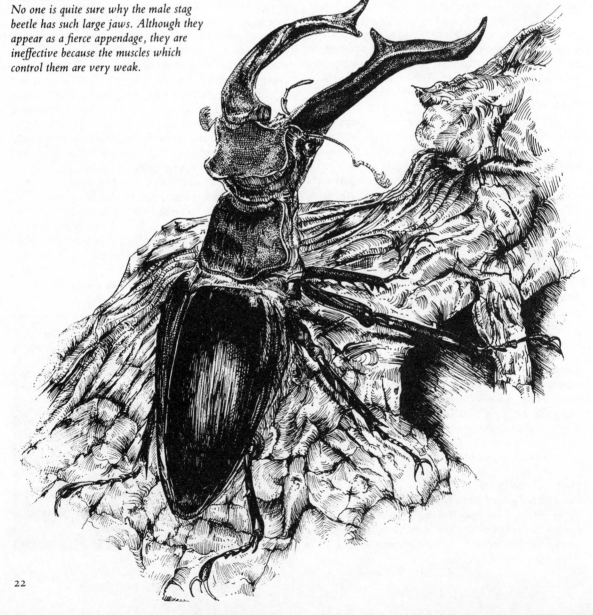

22

Common lizard finds a home

Tips will attract another lizard, a specimen more easily recognised as such. The common lizard (*Lacerta vivipara*), sometimes called the viviparous lizard, because of its habit of giving birth to live young, may be encountered in drier areas of rubbish tips. It is the most common and most widespread of our reptilian species. It may be found in a solitary state, but these lizards normally live in colonies, although these are not distinct family groups and will probably break up from time to time.

At the mating season, which is in spring, the male holds the female on the back, using his mouth to obtain a firm grip. After mating, as with the slow worm, the eggs develop inside the body of the female. The young emerge from the egg whilst still inside the female, and so are born live. Before the actual birth, the female will

---

COMMON BLUE BUTTERFLY
(*Polyommatus icarus*)
On many areas of waste ground there may be plants belonging to the pea family, including birds-foot trefoil (*Lotus corniculatus*) and clover (*Trifolium repens*). The common blue butterfly may lay her eggs on these plants, and some other members of the pea family. She places them on the upper surface of the leaves, and they will hatch in six or seven days. The caterpillars will be fully grown in forty-five days. Some will pupate; others may still feed, and then spend the winter in hibernation.

Common blue butterflies are on the wing from April to August, the adults having a life span of between two and three weeks.

---

make a small depression, which will temporarily accommodate the young, or on an odd occasion the eggs, when these are laid before the young have hatched. Between three and ten young will be produced in August.

At birth the young lizards are no more than 2.5cm (1in) in length, although they are minute replicas of the adults. They must capture their own food at the onset, and will take only very small invertebrate animals. Once they are born, the female does not appear to take any further interest in the young.

A characteristic feature of the common lizard, as with all British reptiles and amphibians, is that the skin is shed at regular intervals.

As winter approaches the common lizard actively seeks out suitable hibernatory quarters. In some instances the reptile may hibernate on its own; in other circumstances there may be small colonies.

Small species abound

By far the largest numbers of species likely to be found are the invertebrates, but the variety is so great that it is only possible to look at a small selection of the more interesting ones.

Because cockroaches (*Blatta orientalis*) will feed on almost any type of material, be it of plant or animal origin, the rubbish tip can provide a great deal of food, and colonies may become established there from time to time. Cockroaches have not only made use of refuse tips for food, breeding and shelter, but some have actually been discovered on the tube – riding around the Underground in London, courtesy of the Central Line!

In Britain there are three species of cockroach, but generally we pass them by, partly because they are not particularly large. It is the species which have been brought to our shores accidentally which are the ones we normally come across, mainly because of their size. The common cockroach does not have the power of flight, but it does have a fast turn of speed. Although they may be seen during the day, the majority are nocturnal species.

Once the eggs have been laid the female carries them around in cases which can be seen hanging from her abdomen. After a few days she disposes of them by placing them either on or in the soil, and the young take care of themselves as soon as they hatch.

Musical insect

The grasshopper is well-known for the 'musical' sounds it makes, which is produced when it rubs the upper part of the hind legs on its wing covers. This insect has a rival in the house cricket (*Acheta domesticus*), of which the male has the ability to 'chirp'.

Rubbish tips provide ideal situations for crickets, since they need the warmth provided by fermenting materials on the tip – even in the winter there is still warmth in the habitat, and, just as important, plenty of food. Although generally termed vegetarian, the common house cricket is cosmopolitan in its tastes. It may eat newspaper, plant material or the remains of animals. Like cockroaches, the cricket may be out during the hours of daylight, but its greatest period of activity is at night.

It is not a native, but arrived here some 400 years ago. Thought to have originated in south-west Asia and north Africa, aided by man it has been dispersed to almost every part of the globe. The sound which the male cricket makes is important during the breeding season. Ensconced in his burrow, he will chirp by rubbing his right forewing against the left one. Guided by his 'song' the female will eventually discover the musician. Crickets have ears, or perhaps 'hearing organs' would be more accurate; they are to be found in the front legs. To lay her eggs the female is equipped with an ovipositor, a tube down which the eggs pass.

Troublesome insects

We will possibly come across gnats, mosquitoes and midges frequenting rubbish tips. Biting insects like the gnats may be extremely troublesome in a 'good' season. The great period of activity for gnats – and mosquitoes as well – is towards sunset and a few hours after sunrise. The female helps herself to a blood meal before the eggs are laid. She may also need nourishment after mating.

Generally speaking their life is spent in the great outdoors, though during the autumn and winter the insects may come into houses when seeking out hibernatory quarters.

While we generally associate the winter months with a lack of insect activity – and for the most part this is true – there are some species which can tolerate even these adverse conditions. Amongst those to be found on the wing at this time of the year are various species of midges and gnats. Of course they are not found when the weather is very severe, but they do appear on winter days.

In most people's minds mosquitoes are associated with the tropics, where they seek their vengeance on any human silly enough to go to bed without some form of protective netting. Mosquitoes occur in many parts of the world, however, and our rubbish tips, amongst other sites, prove a useful breeding place. The most common species is *Culex pipiens*, usually referred to as the common gnat.

Areas of water are often found on rubbish tips. Sometimes these are large where rainwater has collected; sometimes they simply consist of a tinful. Even the smallest water area is a suitable site for the female seeking out a place to lay her eggs. These eggs are cemented together, and in this way they form a miniature raft, which is buoyant enough to float.

When the eggs are ready to hatch, the larvae release themselves by emerging from the part of the egg under water, and become attached to the surface film, their curled bodies suspended here. Although they have need for the water, they also need air to breathe, and can only do this by pushing their respiratory tubes through the surface. They are likely to swim about under the surface, but will return to the surface regularly for a continuing supply of air. The larvae find enough food in most water, for even the smallest puddle will provide the developing insects with minute aquatic living organisms.

We generally associate mosquitoes with the bites which they inflict, though some never attack man, the

female taking her nourishment from birds. However, the largest species to be found in Britain, *Theobaldia annulata*, will take its fill, thinking nothing of driving its mouthparts through even thick clothing, as it searches for a meal of blood.

Theobaldia annulata is one of the aptly termed winter gnats, which will generally brave the elements. The insects are large and grey, with white rings on their legs. Having been fertilised before the bad weather set in, the pregnant females will search out suitable winter quarters, and these are likely to be away from the rubbish tip.

Worms on the tip

The earthworm is described in the garden chapter, but there are other species of worm which might be discovered at the rubbish tip. Unrelated to the more familiar earthworms is one group of animals, collectively termed nematodes or roundworms. Soil scientists have calculated that these species occur in every conceivable habitat throughout the world. Thus these are considered more abundant than the very prolific insects.

Nematodes need another animal or plant as their host, for they are parasites. It has been suggested that every species of plant and animal, including man, on the earth is host to one or other of the many species of roundworms which occur. It does not matter what habitat an animal or plant may live in, as the nematodes have invaded land and water.

You have probably never seen a nematode worm. Most, you will probably be pleased to hear, are exceedingly small, perhaps under a millimetre in length. This accounts for the fact that almost 100,000

CINNABAR MOTH (above)
The attractive female cinnabar moth will lay her eggs on ragwort plants. These will provide food for the caterpillars when they hatch.banding.

SIX SPOT BURNET (left)
The burnet moths are day-flying insects. The female six spot burnet lays her eggs on clover, which provide the caterpillars with food when they hatch.

worms could be found in a single decaying apple, if scientists have got their sums right.

The decaying material in the refuse tip will provide an abundant supply of food for many species which go about their tasks unnoticed. There are some, however, which can be seen, and do emerge from time to time. Living in the soil is a group of roundworms, generally termed thunderworms because they are particularly active after heavy summer (thundery) showers. Brown, and barely thicker than a strand of thread, these worms will sometimes leave their underground haunts, and can be seen making their way slowly over rain-soaked plants. Their exposure is usually short-lived, and as soon as the sun starts to dry out the plants, the thunderworms make their way back into the soil.

Before it lived an independent life it relied very much on other animals. The eggs are deposited in the soil, and when the thunderworms first hatch they attach themselves to, and then push their way into, the bodies of various insect species, including beetles. Until they are fully grown, they will get all the nourishment they need from the host insect.

Flightless insects

Although many insects have evolved wings and attained the power of flight, many flightless primitive insects still survive. Given an abundance of dead and decaying vegetable matter, innumerable springtails (*Collembola*) will occur. They become particularly noticeable when rubbish or damp material is moved. The majority of springtails live unnoticed in the soil, not very far below the surface. So great is the population that it has been estimated that, where they are at their most prolific, there may be over two million of them to each square metre.

Small, the largest generally being slightly in excess of 5mm (0.2in) long, they have acquired their name 'springtails' from their ability to spring. Their forked spring is positioned towards the hind end of the body. Here it is attached by means of a 'hinge' which is dropped down when the animal wants to move suddenly.

Shell-less invertebrates

Parts of the rubbish tip will usually provide ideal habitats for those apparently shell-less invertebrates, only able to survive where moist conditions persist, the slugs. Where they occur in the garden slugs are credited with doing a great deal of damage to plants, and so are relentlessly exterminated. True, there are the destructive species, but not all slugs should be labelled in the same way. No-one is actually likely to grumble at their presence on the tip – better here than in the allotment! Dead leaves and other similar material will provide the slugs with their food supply.

Although slugs and snails are viewed as being different by most people, they are related. The apparently shell-less slugs are not completely without a shell. There are those which still possess some vestige of their former 'home', a feature which snails have managed to retain. Lacking the ability to withdraw into a shell, they are even more vulnerable than snails. Not only are they more easily affected by the elements but they are also likely to be attacked by enemies. Yet, despite their vulnerability, only in exceptional prolonged spells of low temperatures will they not be found. When weather conditions are bad, the slugs burrow deeply into the soil for protection.

As with many species slugs are nocturnal. However, one condition draws them out in large numbers during the day. After a good shower of rain, and with the vegetation dripping with water, slugs will often come out to feed.

As slugs breathe through the skin, as well as through visible breathing holes, the whole body is covered with a slimy material called mucus which helps with the exchange of gases through the skin. They have four tentacles, and the larger of the two pairs have eyes at the end.

The slug is host to a number of visitors. Small mites almost continuously parade over its surface, the sticky mucus seeming to do little to impede their progress. Indeed, it is thought that the mites probably feed on the mucus.

Burying the dead

The sexton beetle (*Necrophorus vespillo*) and burying beetle (*Necrophorus humator*) will be around the refuse tip to 'bury the dead'. The insects work hard together in pairs, burying dead birds and mammals. With very powerful leg action each beetle digs away the soil, and within quite a short time the body of the dead creature will sink into a hole. The female beetle will then lay her eggs in the corpse. When the larvae first hatch they will be fed on regurgitated carrion from the parents, but as they grow they will be able to take their own food from the corpse, which will provide a plentiful supply of food.

These beetles do not spend all their time below the body of an animal. They have the power of flight, and sexton beetles in particular can travel quite long distances and are attracted to bright lights. The insect itself is quite large, and those measuring more than 3.75cm (1.5in) are not uncommon.

A frequent visitor
As with many other areas in the town, a whole variety of mammals may visit the rubbish tip, and perhaps even manage to live on the top. Both species of rat – the ship (or black) (*Rattus rattus*) and the common (or brown) (*Rattus norvegicus*) – are likely to find shelter and food. Although rats are so common, neither species is native to our shores. Ever eager to take advantage of man as he colonised new areas or discovered new shores, the rats followed him. Almost as soon as man had transport, the rats took advantage of a free ride. The first to arrive in Britain was the ship or black rat. Although we are not quite sure exactly when this species reached our shores, it was sometime during the twelfth or thirteenth centuries. It was particularly active around quaysides of the time, and could easily make the leap from shore to ship and back to shore. Perhaps it came with the Crusaders returning from Palestine.

Even today the black rat is found mostly around ports. New rats arrive from time to time on foreign ships and quickly establish themselves on shore. By interbreeding with the established stock, they will produce stronger strains.

By far the most common species, and one which lives and breeds regularly on rubbish tips, is the brown or sewer rat. It is also known as the common rat and the grey rat. It did not come to Britain until the eighteenth century, but has now become a serious pest, consuming vast quantities of food and contaminating just as much more, making it unfit for human consumption.

A native of Asia, when the brown rat arrived in Britain its food consisted mainly of grain. However, as its numbers increased it found that the amount of food available did not satisfy its needs. It is now best described as a scavenger, feeding on both plant and animal materials – in fact almost anything it can find.

Large numbers may occur in a single rubbish tip, and in extreme cases populations of more than 250

pairs may be found. The number of young born to this size of colony can run into thousands in one season.

The rat is capable of tunnelling, and its nesting site is usually below ground level. It is not unusual to find a number of runways leading into the breeding nest. Mating may take place frequently from spring to autumn. After a twenty-four day gestation period, a litter of between five and twenty naked nestlings will be born. Depending on conditions, including weather and food supply, there may be up to seven litters a year. Once fully grown, the brown rat can expect to live for three to four years.

Living in buildings
Around most areas of waste ground there are usually some buildings. Some of these may be inhabited, others deserted. Wildlife is so adaptable that it will take up residence at every opportunity. Indeed there are many species of both plants and animals which manage to establish themselves. When the buildings are inhabited most of the invaders will be short-term

1 YEW
Most yew trees are either male or female, although there are some which carry flowers of both sexes. The use of the wood has changed over the centuries: at one time the springy nature of the timber made yew ideal for longbows, but nowadays its attractive grain can be seen exposed on high-quality furniture and on house signs.

2 HOLLY
At one time it was thought that the leaves on the lower branches of the holly which bore spines were a protective feature. Now naturalists think it is one of nature's ways of ensuring the conservation of water in difficult conditions.

3 CYPRESS
This tree is often grown in towns because of its quick-growing nature. In older trees the bark of the lawson's cypress may be up to 2.5cm (1in) thick.

4 SCOTS PINE
The Scots pine has the distinction of being the only native British conifer; all other species have been imported. It takes the female cone two years to ripen, being green and quite small in the first year, but becoming the familiar brown in the second season.

5 NORWAY SPRUCE
Although the Norway spruce will grow well on poor soils, it has a shallow rooting system, and high winds often bring it down. The tree has an affectionate place in many homes: it is the 'traditional' Christmas tree.

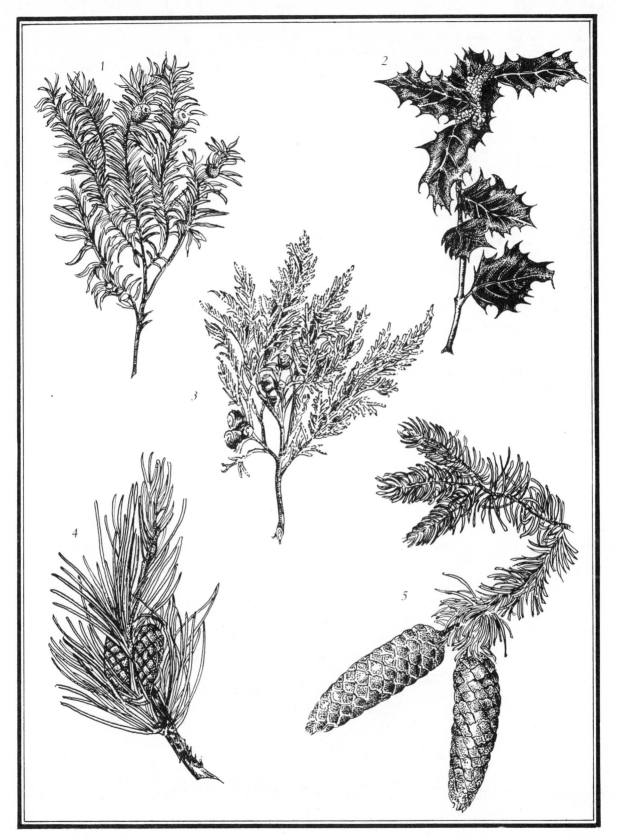

lodgers, and when they are found we tend to evict them. Others manage to live for a long time, perhaps for years, before they are discovered. This is especially true once the buildings become uninhabited. Temporary residents may include small mammals, such as mice. Long-term residents may include such species as woodworm and dry rot.

At one time the house mouse had a relatively insignificant existence, confined to the south-western areas of Asia. Now, thanks to a great deal of unplanned help from man, it is found in every continent. The wary house mouse will make its quarters close to, but usually out of sight of, man. Areas beneath floors, in lofts and cavity walls, will all provide the animals with a home. Generally speaking, they will stay hidden during the day, coming out to feed at night. Although breeding is often confined to specific times of the year, where the mouse lives indoors conditions are much more favourable and reproduction takes place throughout the year. In this sort of environment up to *ten* litters may be produced; the average is more likely to be five or six. The nest consists of shredded materials. What is used will depend on what is available, and the mouse makes use of what it can find.

Feeding on a wide variety of materials, the house mouse is truly omnivorous. It eats almost anything, but seems to be able to thrive on products as diverse as cheese and household soap!

Accidental invaders

Some animals which find their way into houses often do so accidentally, maybe attracted by sweet liquids. Unfortunately, many species such as wasps and bees, if annoyed, may sting; and there are other much more irritating species (much more numerous too) which bite. The sting which the wasp has evolved is not specifically for dealing with humans. Its real use is for dealing with living prey, when the insect uses its sting to immobilise it. Bees (honey bees, bumble bees and other solitary bees) all possess stings which act as defence mechanisms. The larger bumble bee has a particularly powerful sting. Fortunately, it will not use this immediately and has to be particularly angry before doing so.

Bloodsuckers

You only have to mention the word 'flea' to some people and they start scratching – how many readers are at it already? In spite of our claim to be extremely hygienic in the 20th century, our modern society, with its multitude of pets, has seen an increase in the numbers of fleas and lice. Two species in particular cause annoyance to humans. These are the head louse and the body louse. They feed by sucking blood, but they are not the only species which feed in this way. The dog louse is another species which may find a temporary home on the body of man.

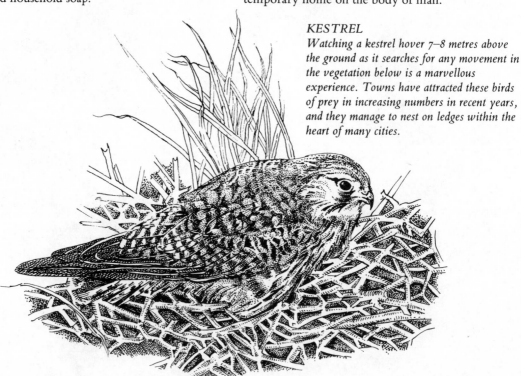

KESTREL
Watching a kestrel hover 7–8 metres above the ground as it searches for any movement in the vegetation below is a marvellous experience. Towns have attracted these birds of prey in increasing numbers in recent years, and they manage to nest on ledges within the heart of many cities.

There are few species of sucking lice, and in order to survive they have evolved specialised techniques for obtaining food. These thrive where conditions are right (very close to the surface of the skin) and this is true of both human and dog lice. It is from the skin that they will extract their two daily meals. Lice need to feed regularly, and when full of blood the body becomes distended. If they are removed from their normal living situation they will probably die very quickly. The temperature of the body is very important to the louse. If it becomes too cold, usually at death, or too hot, in fever conditions, the louse will not be able to live and moves to another suitable host.

The life span is short, the adults only living for a period of 30 days or so. During that time the female will do her best to ensure the continuance of her own kind, by laying some 30 eggs at the rate of 10 a day. The task of removing the eggs from the human body once they have been laid is not an easy one. Both adult and egg have special adaptations to ensure that they have a tight grip on human hair.

Aphids

There is nothing worse than to bring in a bunch of flowers from outside only to find that they are infested with greenfly. Not only do these small insects attack cultivated plants, but they also feed on a wide variety of wild ones – many of which may be outside the house on waste places.

The population of aphids is often controlled by the preceding winter's temperatures: when conditions are mild many will survive, but severe winters will drastically decrease, if not totally eradicate, the population. A similar situation applies to one of the greenfly's natural controllers – the ladybird. In mild winters large numbers survive, and even in severe conditions there are still some which manage to withstand the rigours of continuously cold and wet weather.

Aphids feed by sinking their piercing mouthparts into the stems of plants, and withdrawing sap, which provides them with a source of food. In addition to this it is also known that they carry various diseases from one plant to another.

As well as those which survive the winter, the adults have laid eggs in the previous autumn, and when the temperature rises these hatch out. From these eggs emerge wingless females. Within a short time these too are increasing the aphid population by producing a new generation. The females give birth to live young. At this stage they do not need to mate, and so males are superfluous to their activities.

Some of those which appear from these births are winged females, and in this way the insects are able to move to pastures new. Here they soon repeat the life cycle. It is not until much later in the season that the males appear. These are necessary to fertilise females which then lay eggs for the following spring.

Hoverflies

Another distinctive group of insects which find their way indoors and are much maligned are the hoverflies. Boldly marked with yellow and black, at first sight they are often mistaken for wasps. Nature has decreed that by adopting such a strategy they will be afforded some protection from would-be predators. In the early stages of life it seems that birds, for example, endeavour to eat all kinds of insects. Within a short while they find out which are palatable and which are not. Thus, finding that wasps are not good to eat, they must identify all similarly marked and coloured insects with an unpleasant experience, and so leave them alone.

Hoverflies are aptly named, and can be distinguished from wasps by their ability to hover. Indeed this attribute enables them to remain virtually motionless in one place for several minutes should the need arise. All these hovering insects are grouped together under the heading 'hoverflies', but there are many different species, which are generally only identified by their Latin names. One of those most frequently encountered is *Syrphus ribesii*. This species visits flowers to extract the nectar which it needs for food, but it often rests awhile on the leaves, especially on a sunny day.

ALONG THE LINE

The railway embankment

Few people could have realised the impact railways would have on the countryside. Following the work of the railway pioneers, notably Stephenson and his Rocket in 1841, large tracts of often wild countryside disappeared as mile upon mile of metal was laid down in uninteresting parallel lines, snaking across the countryside until, ultimately, they connected almost every human habitation.

As the great era of the railways reached its peak, a phenomenal amount of countryside had vanished. Now, because of the effect of the Beeching axe, perhaps naturalists should be thankful for the chop which came to many miles of railway lines. Many unused lines have become important sanctuaries for wildlife, particularly in relatively inaccessible areas. As well as taking away wildlife habitats, however, the

WREN
Although a relatively common species, the small brown wren is more often heard than seen. Its delightful song is extremely loud for such a small bird.

BUMBLE BEE
Having spent the winter hibernating in some suitable place, during the winter, bumble bees come out in spring, and almost lazily drone around the flowers in search of their next meal.

WORKER BEE
The worker bee is out doing his job, as its name suggests. In search of food to feed to the grubs in the hive it will spend much of its time labouring over its chores.

BRAMBLE
The white or pale pink flowers of the bramble entice many an insect to taste the sweet nectar. In doing this they will help to bring about pollination so that the familiar blackberries will appear in the autumn.

CONVOLVULUS
Needing other plants for support, the convolvulus will climb upwards in its search for light. Its large white flowers are a joy to behold.

railway did eventually provide new, if somewhat different ones, and operating railway lines and embankments, particularly in urban areas, have provided a protected place for wildlife to live.

Plants everywhere

A wide range of plants, from the smallest lichens to the larger trees, may be found along the railway embankment. On embankments leading into, and out of, urban areas, plants which will not be encountered elsewhere in the urban environment manage to flourish. Once a plant gains a foothold it is likely to increase its distribution along the tracks.

There is only one word to describe the bramble (*Rubus* spp.) – prolific. The bramble, or blackberry, as it is most popularly known, abounds in this waste land. Although it is difficult for the layman to distinguish one species of blackberry from another, the dedicated botanist knows his brambles. No fewer than four hundred different varieties have been discovered in the British Isles! It is quite likely that there are many more awaiting the attention of the experts.

To the country-dweller the blackberry is important because it provides a free supply of fruit in the autumn. This is not strictly true in towns, where access to railway embankments is generally limited; but the brambles will provide a valuable source of food for several different species of animals.

Once established, the bramble is almost impossible to eradicate, and it does not take it long to colonise large areas. Beneath the soil an extremely thick and active root stock seems to send up a continuous supply of stems. Once above the ground these grow quickly, and eventually bend over so that the end touches the soil. When this happens, the stem tip-roots. A new plant will become established when the stem is severed from the main plant.

When the plant increases in this way, it is not very long before the bramble takes over large areas, forming a dense and, because of its effective defence mechanisms of spines, prickles and hooks, impenetrable scrub.

Although the plant is deciduous, it is not unusual for the leaves to remain on the blackberry throughout the winter. Indeed, various beautifully coloured tints can be found for much of the winter and spring, until young new leaves are produced.

Insect visitors and pollination

With so many varieties, the range of flowers is wide, varying from white through pink to pale purple. The first bramble flowers appear in small numbers, with the greatest show of blossoms occurring in June and July. In most years it is quite normal to see flowers persisting on the plant well into the autumn.

Producing plenty of nectar, the flowers will attract innumerable species of insects out in search of a rich supply of nutritious food, and bringing about pollination. The largest visitors will undoubtedly be the butterflies, and once they settle on a flower their wings will effectively cover many more. Having imbibed as much nectar as possible, they will move on to others until their appetite is satiated.

In inaccessible places the berries that the pollinated flowers produce will usually not attract the hand of man, and so will provide a valuable and almost unending source of food for a whole host of invertebrates. The outer covering of the blackberry is tough, and many insect visitors are unable to tear this apart to get at the succulent flesh inside; but wasps, their powerful mandibles performing the task, will open up the fruits for others to enjoy. The insects will not be the only beneficiaries to share in the bountiful harvest. Small mammals will be able to take their fill, especially those, like the mice, which are capable of climbing. Fruits will fall to the ground, and these will be eagerly

> **ADDER** (*Vipera berus berus*)
> The Latin name *vipera* gives the adder its common name of viper. It occurs from time to time on dry railway embankments, especially if there are spots where it can sun itself.
>
> It is our most common snake, and will be found in most areas. In some places it seems to live quite close to man; in others it does not occur. Naturalists suggest that it does not like disturbance. It is almost certainly most common where it is least disturbed.
>
> In the spring the males assert their right to their own territory by coming together, and performing what has come to be known as the 'adder-dance'. Males lift up their bodies in an effort to drive out the weaker ones.
>
> After mating, in which the male holds the female in his mouth, she will eventually produce between five and eighteen young.

seized by mammals which lack the ability to negotiate the tricky bramble stems. Hedgehogs and foxes will probably stop by for a tasty morsel.

A climbing plant

Where the railway embankments are dry and grassy, with a sandy soil, the corn (or field) bindweed (*Convolvulus arvensis*) will probably take root. Although it causes the gardener much anguish as he tries to get rid of it, seemingly never quite able to dispose of all the roots, where it grows on waste ground it is much more likely to be admired.

Convolvulus is a perennial species, and its aerial parts die down during the winter. The plant has a continually spreading rootstock beneath the ground, enabling it to colonise new areas. Above the soil the root sends up an almost unending array of shoots, which in turn give rise to a large number of twining stems. The spear-shaped leaves are arranged alternately along the stem.

The flowers are particularly attractive when fully open in sunny weather. Varying in shade from delicate pink to white they are usually subtly but distinctly streaked with pink. Anyone attempting to pick the cup-shaped flowers may find that they close up. The

HEDGEHOG
The unmistakable black, almost cigar-shaped droppings are evidence that the hedgehog has been in gardens, which it frequents. There is so much 'wild' land in towns, however, with suitable daytime resting sites, that the hedgehog has a wide distribution. It is valuable as a pest control agent, and charges the town-dweller nothing for its services.

plant reacts in a similar way if it rains, and at night, a mechanism which conserves the quality of the nectar, as its flowers are folded in neat pleats.

Although no more than four seeds are produced by each flower, they are equipped with a tough outer casing, which enables them to withstand extremes of temperature, trampling and so on. Thus they can remain viable for several years.

An established foreigner

A plant which has done particularly well in Britain considering that it is a native of warmer climes, is the Oxford ragwort (*Senecio squalidus*). Its original home is far from this well-known university city, although it does owe its British origins to Oxford. Related to the dandelion, it is a native of the island of Sicily and the Italian mainland, favouring volcanic soils; but unlike some other species it did not reach Britain accidentally. Towards the end of the eighteenth century, botanists collected the seeds from its native land, and grew them in the Physick Gardens at Oxford.

It did well in its new home, and produced good crops of fertile seeds. If botanists thought that they could contain it within the confines of the garden, they were mistaken. The light, ripe seeds left the parent plants and were transported to pastures new by favourable wind currents.

The plant was first noticed outside the garden growing on some old walls in the university city, and seeds were carried to new areas in the normal way; but the next part of the story of its dispersal is more exciting. The even greater and more widespread distribution of the plant was due to the coming of the railways. Seeds fell on the clinker which filled the spaces between the railway sleepers. Such an area seemed to be an ideal habitat, akin to the volcanic soils from which the plant had been wrenched. Man, or perhaps more accurately, machine, did an effective job as the trains moved to and fro over the tracks. The light

extend from May through to December. It often brightens up railway track sides with a splash of colour when most other flowers have faded.

Another ragwort

Along some areas of railway tracks, the bright yellow flowers of the common ragwort (*Senecio jacobaea*) are particularly striking. A member of the daisy family, generally common over much of Britain, its flower heads are daisy-like in structure. A perennial, the plant may be found flowering from June to October in many areas, and in the south it may manage to bloom as late as December.

The flower heads are arranged in more closely branched clusters than in the Oxford ragwort. There is a considerable difference in height between the two ragworts. Oxford ragwort seldom reaches more than 30cm (12in) in height: common ragwort varies between 60 and 90cm (24 and 36in). The leaves of Oxford

---

HAIRY BITTERCRESS (*Cardamine hirsuta*)
In any habitat the survival and germination of seeds is of vital importance if the plant is to survive. Those which fall to the ground underneath the plant have much less chance of becoming established than those which are dispersed away from the parent plant.

One such is hairy bittercress. It belongs to the group of plants which, because of their particular attribute, have become known as 'touch me nots'. They only have to be touched when the seeds are ripe and they are ejected.

The seeds are protected inside a pod. As they ripen, the pod becomes dry and taut. If the plant is touched in this state, then the two parts of the seed case twist back and the seeds are fired away. Fairly light, they will be carried by the wind, and even when there is hardly a breeze they may be thrown as much as one metre (39in) from the parent.

GRASSES
Grasses are generally particularly adaptable and are found in many habitats. Propagation is both by underground rhizomes and by seeds. The flowers of grasses are generally overlooked because of their drab nature.
1 Creeping bent grows where the soil is rich. A species found in fresh water marshy areas, it has been added to lawn grass mixtures because it produces an extremely durable sward.
2 Yorkshire fog may be found in habitats as wide apart as damp meadows and roadside verges. It is therefore just as likely to be encountered in urban environments as in the countryside.
3 Cocksfoot has shoots which, characteristically, stay green even throughout the harshest conditions. Once spring arrives it will soon begin to grow strongly.
4 Hare's tail is a cotton grass which is found in various parts of the British Isles; it can be encountered in urban conditions where it grows in marshy places.
5 Timothy grass derives its name from an American, Timothy Hansen. It is now widespread in many habitats, although it probably originated in low-lying grassland and in wet meadows.
6 Wall barley becomes established where there is either disturbed soil or waste ground and fares particularly well when it is close to buildings and walls around houses. Flowers can be found on the grass from May until August.
7 Perennial rye grass is found in many situations, because it can withstand constant trampling, and is especially tolerant of these conditions.
8 Barren brome will be found as a well established species where there is waste ground. It will be encountered by the roadside, in disturbed areas and along railway tracks, amongst other habitats.

seeds were probably drawn along by the air behind fast-moving trains, resulting in the rapid spread of seeds in some areas. Soon ragwort plants spread over a much wider area, reaching the capital city in the nineteenth century. As the species slowly moved out from Oxford, it ultimately reached Scotland, being recorded there in 1940. Now found beside railway tracks over much of England, the plant has also become established in other habitats.

With composite yellow flowers, the Oxford ragwort may succeed as an annual, a biennial and sometimes as a short-lived perennial. It has a long flowering period which, in favourable seasons, may

ragwort are bright green: those of common ragwort are duller.

According to folklore, fairies in the highlands and islands used stems of the ragweed – as it was popularly called – to propel them from one island to another. The plant was important to them in wet weather as well, since it afforded them protection: they would shelter under its leaves.

The plant gets its common names of ragwort and ragweed because of its 'tattered' leaves. In Scotland, it was called stinking Willie because the horses of William, Duke of Cumberland, were responsible for spreading the seeds during the Battle of Culloden. In earlier times it was known as *herba santi jacobi,* St. James'

SONG THRUSH
Attracted to a wide variety of habitats in the town, the song thrush is very popular because of its beautiful song. Fluent throughout the year, with the exception of August, the bird repeats its notes twice and sometimes more.

herb, since it was generally thought to be in flower on the feast day of St. James, 25 July.

The ragwort's guests
Strikingly coloured, like the plant, the cinnabar caterpillars feed voraciously on suitable ragworts. In their eagerness to take every available piece of leaf, they will quickly clear everything except the stems. Where they occur, each ragwort plant will usually house large numbers of cinnabar moth caterpillars, since they are gregarious.

A distinctly conspicuous species, the cinnabar moth could be considered an 'exotic' species at first sight. With its deep rich pink hind wings, it will be on the wing in May, and can be found over much of Britain. Sometimes a specimen with yellow, instead of the more familiar pink, hind wings is encountered.

GOLDCREST
In spite of its diminutive size, the goldcrest has a reputation for being a fighter. Cock birds at least will become involved in fights when a hen bird is at stake in the mating season.

The common 'weeds'

Many plants normally termed common weeds will usually be found somewhere along the railway track or embankment. One group which manages to do quite well is that of the sow thistles (*Sonchus* spp.), plants belonging to the Compositae, a group which encompasses many other composite yellow flowering species, like the dandelion. Flowers on the sow thistle are not unlike miniature dandelions. The corn sow thistle is one which, despite its name, will often occur in waste places. The smooth sow thistle, generally an annual, may also be found.

Beneath the ground the rootstock of the perennial corn sow thistle will be continually creeping in a relentless effort to colonise new areas. Runners will grow up to give rise to new plants. Although the sow thistle is generally confined to a maximum height of 1m (39in), where moist conditions are found it can almost double this height. The stems which bear the composite yellow flowers are hollow and contain a white, milk-like liquid, which oozes out when these are broken. The deeply lobed, finely cut leaves have stalks near the base of the plants: those higher up the stem do not possess this feature. There are sharp teeth on the leaves and sticky hairs on the flower stem.

The light, fluffy seeds, akin to those of the dandelion, may be carried over long distances by wind currents before they land in a new area, where they may germinate.

One of the features of the plant is that, even if the flower-bearing stems are chopped down, once the flowers have opened the seed-producing activity continues. Thus cut-down plants can produce as many seeds as those still standing.

Productive shepherd's purse

Come rain or come shine, the shepherd's purse (*Capsella bursa-pastoris*) will produce its seeds. One of our commonest weeds, it seems to find a niche in almost every habitat. Although, as with all flower-bearing plants, it is not unusual to see the small white flowers covered with insects, the plant does not need these visitors to bring about pollination. Indeed, seeds will always be produced, because pollination takes place before the flowers open. This method obviously has a distinct advantage over insect and wind pollinated species.

The plant grows extremely quickly, so much so that in the summer months a plant will have germinated from a seed, matured and produced its own seeds, all of which has taken place within the short space of six weeks.

Prolific throughout most of the year shepherd's purse can often be seen in flower in winter. Not having to rely on insect pollinators – few and far between at this time of the year – the plant may scatter its seeds even in conditions which many other plants find totally intolerable. With each plant having a short life cycle, large numbers of seeds will be produced annually. Bearing in mind maximum and minimum growth periods, there may be as many as four generations in a single year. With each plant producing in excess of 2,000 seeds, it is not surprising that the species is successful. Although most of the seeds will begin to germinate almost as soon as they reach the ground, in unfavourable conditions they do not, and may remain dormant and viable for several years. They can be transported over large areas by both birds and mammals. When the seeds become wet they have a

RED ADMIRAL

Our damp, cold winters generally defeat the red admiral's attempts to hibernate, but many arrive annually in our country from the Mediterranean. Once they have mated, the female lays her eggs on stinging nettle leaves. Red admirals are frequent visitors to gardens in autumn, when the adults seek out the nectar from plants such as michaelmas daisies.

slightly tacky texture, and adhere to feet and perhaps clothes. They will also pass through the bodies of birds without being digested, and when they are expelled in droppings – perhaps many miles from where they were picked up – they may start to grow.

There is a rather wicked little game which children used to play. Once the seed cases have formed they look like miniature hearts, and hence their alternative name 'mother's heart' is appropriate. The seed case is extremely brittle when ripe, and if pulled from the plant will break in two. In the game one child bids another to pull off a seed pod, and then when he does this, and it breaks in two, the youngster claims that the offender has 'broken his mother's heart'.

The seed case resembles the sort of purse which was used in earlier times. According to some stories, a small purse on this almost insignificant plant was compared to a similar container of coins on a poor man – a shepherd. Hence the plant's common name.

The scarlet heads of poppies

Although the common red or field poppy (*Papaver rhoeas*) is usually considered a plant which is found on cultivated land, it also manages to do well in waste places, like railway edges and embankments. The small seeds are easily transported from one place to another. In fact there are two very closely related species: the one already mentioned and the long-headed poppy. The differences can mainly be seen in the shape of the seed capsule. In the long-headed poppy the capsule is elongated, as the name implies. In the field poppy it is stouter.

The brilliant red petals will attract insects, and although they do not produce nectar there is an abundance of pollen, which bees in particular will relish. In this way these insects bring about pollination by taking pollen from one flower to another. The insects must visit the poppy flowers quickly, for once they are open the petals may fade and fall the same day.

PEACOCK

It is the conspicuous eye-spot on the peacock's wings which give this butterfly its common name. Like the red admiral and the small tortoiseshell, it lays its eggs on the leaves of stinging nettles. Like them, it also enters gardens, especially in late summer and early autumn, to drink nectar from flowering plants.

Fertilisation completed, the once-insignificant capsule will grow very quickly; when it is fully ripe there are a number of slits around the top under the over-hanging rim. Even the slightest breeze will make the poppy stem sway backwards and forwards, which scatters the seeds.

Whereas the seeds of the field poppy will remain dormant throughout the winter months, those of the long-headed poppy (*Papaver dubium*) start their growth soon after they have been liberated. The outer seed coat of the field poppy is tough, enabling the seeds to remain dormant in the soil not just for one year but up to a hundred years, according to some experts.

Inside the two protective bristly sepals, the poppy petals look crumpled, at first, but the creases soon fall out. The long, slender stem will push the flower high, perhaps almost a metre (39in) from the ground, so that it will stand out like a beacon to passing bees. The four-petalled flower may be up to 10cm (4in) across, and two of the petals are noticeably smaller than the others.

Inconspicuous grasses

A number of species of grass will undoubtedly be discovered along the railway line. These members of the plant kingdom are not always easy to identify, partly because many related species have similar characteristics, which only an expert can distinguish, and partly because many closely related species have inter-bred to form hybrids. One species which is frequently found, especially where there has been some kind of disturbance, is wall barley (*Hordeum murinum*),

which also favours areas close to buildings and walls. The flower head, though smaller, bears a close resemblance to cultivated barley. An annual species, it has a wide range of height: tall specimens may reach 60cm (24in), small ones no more than 6cm (2.5in).

Yorkshire fog (*Holcus lanatus*) is another species which may be encountered. Bearing flowers from May to August, it varies in height from 20 to 100cm (8 to 40in). It is tolerant of a wide variety of soils and so will be found in many areas of the British Isles.

Cocksfoot (*Dactylis glomerata*), so called because the arrangement of the flowers on the stems is a reminder of a cock's foot, will also be found. Left to its own devices, the plant forms a dense tuft.

It is the grasses, along with other species, like the stinging nettles (*Urtica* spp.), which are responsible for the familiar and troublesome hay fever.

Food for butterflies

Where cocksfoot (*Dactylis glomerata*) and the meadow grasses (*Poa* spp.) occur along railway lines, they may provide a supply of food for the caterpillars of the wall brown butterfly (*Lasiommata megera*). The female places the relatively large greenish-white eggs singly on suitable blades of grass. When the caterpillars hatch out, they are extremely conspicuous, the body being covered with dark hairs. However, after the first moult, they take on a more familiar green colour. It is many months before the caterpillar reaches the pupation stage, spending the winter in the larval form. Unlike some species, it will come out and feed during the winter, but it will seldom be seen searching for food at any time of the year. Nocturnal in its habits, during the hours of daylight it remains securely hidden deep amongst the tussocks of grass on which it feeds. The caterpillars which have spent the winter in this state will make the change to adult and be on the wing from mid-May if conditions are favourable. These will mate, and the next brood appears towards the end of July or early in August. It is these caterpillars which remain immature, making the change to the adult winged form the following year. During a prolonged hot summer there may be three broods.

Armed with poison barbs
A plant capable of colonising almost any disturbed area is the stinging nettle (*Urtica dioica*), which provides a valuable source of food for many animal species, including the caterpillars of the red admiral, small tortoiseshell and peacock butterflies. It is the stinging hairs which cause humans the most trouble, but in spite of them the plant can be cooked and is very nutritious, being rich in vitamin C.

According to some of the early folk experts, there was certain to be gold where nettles were found. Although we might despise them today, they were actually grown in some parts of Scandinavia in the eighteenth century, and the fibres used for cloth-making. There is a tradition from Scotland that stinging nettles will grow on the body of the dead. Such a belief seems to have arisen because of the persistent nature of the plant: once established it is difficult to get rid of.

The stinging hairs are to be found on the stems as well as on the leaves. The tip of each stinging hair is extremely brittle, and even the slightest pressure will cause it to snap off. As it does this a very small, fine, needle-like projection is exposed. It is this which enters the skin, and poison is then pumped into the body. This is formic acid, which causes the itching and the small white swellings. Although our ancestors ate the herb, many town people find this difficult to believe. Yet it is still much relished in many rural areas. In fact the poison is destroyed when the nettle is boiled.

Because of their pale colour and almost insignificant nature, nettle flowers are often overlooked. Each plant bears either male or female flowers. In the case of the male ones, these have a distinct, if somewhat drab, greenish tinge, and are long and catkin-like. The female flowers are small and seldom seen. The wind picks up the pollen from the male flowers and carries it along. Some will land on the female flowers and pollinate them. Seeds will be produced, but the stinging nettle also has another means of spreading. The persistent yellowish underground roots send out runners just below the surface. These provide the plant with an effective means of propagation.

Although the stinging hairs cause distress to humans and probably some other animals, the nettles provide a supply of food for some species which feed on the plant quite happily without any ill effects. Below we mention the caterpillars of some butterflies, and there are a number of weevils which eat the leaves. One species of midge brings about the formation of galls.

Common butterflies
Stinging nettles are likely to provide food for the caterpillars of three common British butterflies – the small tortoiseshell (*Aglais urticae*), the red admiral (*Vanessa atalanta*) and the peacock (*Inachis io*). Of these three species, two are native – the peacock and the small tortoiseshell – and the red admiral is a migrant. It is because the fragile insects are equipped with an innate instinct which urges large numbers of them to leave their Mediterranean home for our shores that we have the pleasure of their kind during the summer months.

A native of a warmer clime, they find the conditions during our British winters intolerable. With alternating periods of wet and dry weather, neither adults nor chrysalids are likely to survive to see the following spring. Most will perish as the adverse weather arrives in autumn, although it does seem that some try to make the long migratory journey back to their Mediterranean homeland. The surviving adults which do not attempt the arduous journey will usually seek out somewhere to rest for the winter, but most will have died before the next spring arrives.

As soon as the first migrants reach our shores – usually in May – they will mate, and the females will seek out suitable places for egg-laying. So that the caterpillars can have food from nettles when they hatch, the eggs will be positioned singly on the upper surface of young plants. If older plants were selected they would be tough and could not be tackled by the caterpillars when they hatched out. Once out of the egg-case each caterpillar pulls together the edges of the nettle leaf, binding it with a silken thread which it produces. Only when this is completed does it start to feed on the succulent leaf. Once all possible nourishment has been derived from the food on which it hatched, it will move on. It continues to feed and when fully grown it will bind together a number of leaves to form a tent-like structure. It is here, suspended upside-down, that it will pupate, and make the change to the adult form.

The first British born red admiral butterflies will usually be on the wing in July, and will be ready to breed. Although most Continental migrants arrive earlier in the summer, it is not unusual for them to be

BULLFINCH (above)

In spite of its attractive plumage, the bullfinch has many human enemies. It is the rose-coloured breast which immediately identifies the species. It has a bad habit of taking buds from shrubs and fruit trees. Although it is not a common sight in towns it is encountered in parks and gardens, occurring more often on the edges of built-up areas.

CHAFFINCH (left)

Census figures show that the chaffinch and the blackbird vie for the title of 'Britain's most popular bird', as far as numbers are concerned. It is during the colder months of the year when food is scarce that the chaffinch is typically encountered more often in the town. Then it will come to bird tables and visit parks in search of food. The stout bill of this member of the finch family indicates a seed-eater, but the chaffinch also takes invertebrates during the summer months.

joined by new arrivals right through until September. As long as the females which make land can find a mate, and then find suitable nettle plants, they will continue to lay eggs. Emerging butterflies are likely to be encountered until the end of September or into October.

Woken from a winter's sleep

Small tortoiseshell butterflies spend the winter as sleeping adults. The spring weather will determine when they stir. If it is warm they may be out towards the end of March, and certainly by April. These winter sleepers will mate and be on the wing until May. The female searches out stinging nettle leaves, and she will place between 50 and 100 pale green eggs in a haphazard arrangement on the under-surface of the leaf. As the caterpillars hatch, their first task is to ensure that there is a communal web which covers the food plant, allowing them to feed in relative safety.

As they exhaust the food supply, they will move away, spin a new protective web, and then continue to feed. Not only does the web provide suitable shelter whilst feeding and at night, but the caterpillars also use it as a 'sun bed', taking advantage of the warm rays which penetrate to their home.

Just before the larvae have grown to full size, they leave the company of their relatives and spread out, often moving quite a distance from the original plant. Here they find a suitable leaf, fold it over themselves, secure it with silk, finish feeding, and then usually crawl towards the ground to pupate. Finding a suitable stick in the undergrowth they will hang suspended

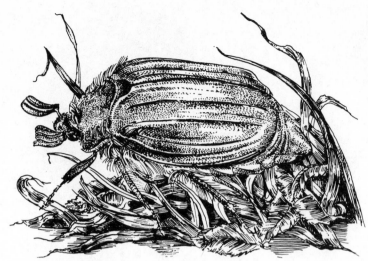

COCKCHAFER/MAYBUG
The cockchafer or maybug is not welcome in most gardens, as it strips the leaves from trees. It is most active in May and into June, hence its common name of maybug. The larvae are active below the ground where they nibble at the roots of many kinds of grasses and cereals.

from this as they change into chrysalids (*pupae*).

The first brood will generally be on the wing in the early or middle part of July. Mating will take place, and they will lay their eggs. The new brood, appearing on the wing in August, may well stay active into October. As the cooler weather arrives, the butterflies will find suitable places to hibernate. The buildings alongside the railway line – and of course far away from it! – will provide them with suitable protective winter quarters.

Identical adults
The third species generally associated with nettles is the distinctively marked peacock. To all intents and purposes, male and female peacocks are identical – at least as far as markings are concerned. It is the springtime courting behaviour which separates the sexes. After they emerge from their winter sleep, the males – now seen to be smaller against the females – will chase their partners in an elaborate courtship display.

HAWTHORN
The snow-white flowers, which open in May, give the hawthorn its popular name of 'may tree'. They are eventually fertilised, and the red berries, the haws, appear in late summer. It is these which are sought out by many seed-eating birds.

HAWTHORN SAWFLY
Sawflies get their name because the females of most species have ovipositors which resemble miniature saws. These effectively make slits in the stems of plants, including hawthorn, and the female then deposits her eggs inside the stem.

COMMON FIELD GRASSHOPPER
(Chorthippus brunneus)

In areas where there is some grass on rough ground, the common field grasshopper may occur.

The hind legs are elongated which enables the insect to jump well. The characteristic of any grasshopper is its ability to sing. In Britain there are twenty-nine different species, each of which can be recognised by its own song. It is the male grasshopper which sings more often. The female usually reserves her song to help attract a mate.

Eggs are laid well below the soil, at the base of the roots of grass. Placed here, encased in a solid jelly-like froth, in summer, they will remain until the next April, when they will hatch. After a number of moults, they will reach the adult stage in July.

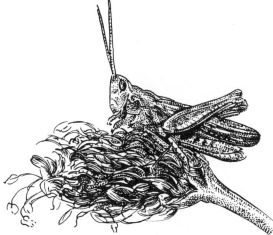

GRASSHOPPER
It is the male grasshopper which generally makes the familiar 'singing' sound. Each species makes a distinctive noise, used to attract a mate. Pesticides and herbicides have caused a decrease in the grasshopper population.

After mating, usually in the early part of May, like the small tortoiseshell, the female peacock will lay her eggs haphazardly on the under-surface of nettle leaves. Once the caterpillars hatch they will live communally. Having spun a web, which in this species is slung between the stem and leaves, they will feed on the leaves. Once they have finished the food supply, they will take their leave, move to a nearby plant, and after spinning a web continue their interrupted feeding. After the final moult before pupation, each caterpillar seeks the company of only a few others, the large family groups splitting up. Before actual pupation, the nettle is abandoned, and the caterpillars seek shelter in nearby undergrowth. Once upside down, they change into a chrysalis before the adults emerge.

On the wing until late into the autumn, their hibernatory quarters will vary according to what is available. Outhouses are suitable, but so are rabbit holes and hollow trees. Disturbed during their winter sleep, they will react by rubbing their wings together. This produces a strange hissing noise, the purpose of which is not known.

Food for man and beast
How useful the elder (*Sambucus nigra*) is to both man and wildlife. Common and widespread throughout much of Britain, it will grow in almost any situation, and flourishes by the side of railway lines. It appears in all sorts of odd places, including walls and buildings. Although this may seem surprising, it is not really, since birds may carry and drop seeds almost anywhere. Where it does not have wide open spaces, it will occur as a shrub. Even in full flight it seldom grows to any great height, probably reaching a maximum of 12m or so (40ft).

As the young tree grows it gives off a rather strange smell, which some people find unpleasant. A similar scent can be obtained by bruising the leaves. The flowers usually appear in June, slightly earlier in the south and later in the north. They are creamy white, and have a pleasant fragrance. Besides providing a supply of food for various insect species, they are collected by some people to make a very fine wine. It is rather later in the year, when the purple berries hang in tight bunches, that both birds and man will be out in search of them. They are a valuable source of food for birds, and much sought after by many folk as a basis for a favourite wine.

Opportunist sycamore
Sycamore (*Acer pseudoplatanus*) seems to do well along the embankment beside the railway and, like the elder, it may also appear in strange places. Walls, the spaces between paving stones, and in gaps against walls – all these sites may provide a growing place for the sycamore. With lack of space to spread, however, it is usually stunted or dies before it reaches any great height. Particularly quick growing, left to its own devices it may reach a height of 30m (100ft) in 100 years; but along the track it will probably never reach these dimensions.

The flowers will be found on the tree in May. They vary in colour from yellowish-green to yellow. Hanging in clusters, they are almost as conspicuous as those of the elder. Unlike the elder, however, they do not possess such a strong scent, yet in spite of this they do attract a large number of flying insects.

In the autumn, the tree will have produced two-winged seeds, called samaras. These wings enable the seeds to be carried by wind currents some distance from the parent plant. It is not surprising, as one watches them almost dancing in the wind, that they have attracted the country name of 'spinning jennies'.

Taking advantage of man
As far as mammals in town are concerned, the great opportunist of our time must surely be the fox (*Vulpes vulpes*). Realising that man is a wasteful creature, he now lives deep within towns, taking some of his

FOX
The fox has become the 'success' story of town wildlife, as a recent 'foxwatch' in the capital city revealed. It is found living in the heart of cities, and may spend the day closer to human dwellings than many people realise. Man has provided the creature with an almost endless supply of food.

nourishment from the dustbin and the garden. Perhaps, too, he has decided to 'stay put' in the areas in which he always lived, which were rural at one time.

The railway line, and particularly the embankment, is likely to afford a suitably deserted area, where the fox can excavate his earth. This may be made under tree roots, in ditches, or he may even take up residence in the now deserted homes of other animals, like rabbits and badgers. In the case of the former some excavation work may be necessary to make the entrance hole larger for easy access. As far as badgers are concerned, it is possible that the fox may live in a disused part of the sett. This is not always the case. In some areas, he may move in with the badgers – although it is likely that they will move away from him.

In ideal conditions the fox is a meat-eater, but often in the urban environment the mammal will have to compromise. Although he will find rats and mice, and perhaps even the odd hare and rabbit, he will have to turn to vegetable matter. Dustbins are upturned in some areas in an effort to take out the remains of chicken and other meat.

Larger dog fox

An adult dog fox will probably measure around 60cm plus 40cm of tail (2ft plus 16in): a vixen will be slightly smaller. Weights average out at about 5 to 7.5kg (12 to 15lb). The fox is a solitary animal, except during the breeding season. At the mating period, the foxes are particularly vocal. The dog fox barks loudly as he endeavours to discover his mate. According to some naturalists, each dog fox has only one particular vixen. Should anything happen to her, it is said that he will lead a solitary life. Other naturalists dismiss this. Breeding takes place in January or February, and the gestation period is usually 51–52 days. There is only one litter in a year, consisting of between three and six cubs. The young are able to see after ten to twelve days.

Whilst the vixen is engaged with her nursery activities it may be the dog fox which will be responsible for hunting. If this is the case he will keep the vixen supplied with adequate food, also supplying the cubs with food before they first leave the earth. In some cases both dog and vixen share in the capture and provision of the food. There are also instances where the fox deserts his mate, leaving her to provide all the nourishment for both herself and her cubs.

Ready to make their first tentative skirmishes with the outside world, but accompanied at first by the ever-watchful parents, the cubs are continually aware of danger. As they gain confidence, they spend a great deal of time at play. Within six months they are fully grown, reaching maturity by the end of the first year. They will be ready to breed at the beginning of the following year, if they are able to attract the attention of a mate.

Mammals find a home

Until the advent of myxomatosis, which started in earnest in Kent in 1953, the rabbit (*Oryctolagus cuniculus*) was a common animal in almost every part of the British Isles. During the onslaught of the virulent disease, it was estimated that no fewer than sixty million rabbits suffered as a result of the epidemic – that is more than the entire human population of the British Isles. Today the once familiar mammal is rare in some areas; in others it is recovering and has become immune to the disease.

The rabbit was supposedly brought to our shores in the twelfth or thirteenth centuries by the Normans, to be bred in areas known as warrens, to provide an alternative food and fur supply. In effect the rabbit was meant to take over from the beaver.

Although the underground homes of some colonies of rabbits may occupy a very large area, in other places the numbers are smaller. Since the onset of myxomatosis, there has also been a tendency for the animal to live above the ground, a feature which was peculiar to the Australian population.

In most areas it is quite normal for some rabbits to breed throughout the year, but in fact the main breeding season spans half the year from January through to the end of June. The greatest period of activity, when the largest number of rabbits will be born, is usually mid-way through the breeding period – March and April. After 28 days gestation, a litter, varying from four to twelve youngsters, will be born in a breeding area – called a stop – off the underground runway system. At birth the young are both blind and naked, and rely on their mother for protection. When she leaves the tunnel, she seals up the stop to prevent would-be attacks. After ten days the young have their eyes open. They can expect to live for between five and eight years.

Since the female will mate some twelve hours after her litter is born, it is not surprising that she gives birth to one litter after another. In this way she can easily cope with between three and five litters in a year.

Along the railway line, and perhaps in other suitable areas in the town, the rabbit will feed on those herbs which are available, supplementing a meagre diet by raiding nearby gardens. In the countryside the rabbit is said to cause many millions of pounds worth of damage to the farming community in the form of lost crops.

Autumn harvest
In addition to the times throughout the year when the flora of the railway embankments is attractive, in the autumn wildlife which either lives there or comes in search of food will usually not go away empty handed. Distinctive berries appear and may stay on the plants for several weeks, depending on weather conditions. Nature, ever versatile, has ensured that the seeds which the birds ate in autumns and winters past will have been dispersed. Provided that there is a suitable niche, many of these will begin to grow, mature and provide food for marauding species.

In a good autumn the elderberries hang ripe and purple, weighing down the branches of the elder shrubs and trees with their harvest. Although called berries, the fruit of the elder is correctly known as a drupe. Opportunists, once an elder seed falls on fertile ground (probably carried there by a bird and deposited

HEMP NETTLE (*Galeopsis tetrahit*)
Common or large hemp nettle is normally associated with cultivated ground, and it sometimes grows in fens and damp woodlands. It is seldom seen in the urban environment, although plants have been discovered on waste ground in Glasgow. The flowers are similar to those of the dead nettle, and they vary in colour, being pink, white or purple. Where suitable conditions favour it, but seldom on waste ground, it may reach a height of nearly a metre (39in).

with its droppings) the seedling develops. Growth continues apace, the stems soon exhibiting a woody nature. In times past they were collected, and the soft central pith removed. The hollow elder stem was then used either as a pea-shooter or as a simple musical instrument. Before the advent of artificial materials, the soft pith was used in the preparation of sections in biological laboratories.

As soon as the elder is capable it will produce its annual display of flowers. Creamy white, they have an abundance of nectar, and will be visited by a variety of insects. Some folks collect the heads in early summer so that they can make elderflower champagne.

Once pollinated the flowers fade fairly quickly. The fruits begin to ripen. At first they are green, but gradually go through several colour changes until they become the familiar purple-black colour. Ripe, there might be great competition for the elderberries. Humans may come in their hordes, filling their buckets and bags. Back home, they turn the drupes into elderberry wine. Birds seek them out too. If you live close to an elder tree, you will know that birds have been eating them when your drive or car shows distinctive purple stains.

Familiar rose hips
Too often we forget how important some of the countryside fruits were in the past. It has been

LONDON ROCKET (*Sisymbrium irio*)
As its name suggests the London rocket was found in the capital. It first appeared, and was discovered in quite large numbers, after the disastrous Great Fire of 1666.

Now a rarer plant, the golden yellow petals still add a touch of almost magic to the desolate scene, just as they did after the fire had ravaged London. A Crucifer—a member of the cabbage family—the flowers are arranged in a cross, the feature which gives this family its name.

It is still found in London and other urban areas.

recognised for a long time that the fruit of the wild rose – the hips – are rich in Vitamin C. Many folk must have had experience of collecting them in the last war, when they were sent off to factories to be used to produce the famous rose hip syrup.

Wild roses abound along the railway line. The plant, the ancestor of all our modern roses, is attractive in many seasons. The flowers of the wild rose may be pink or white. Whatever colour they are, however, they will always be five-petalled. Sweet-smelling, the plants will attract the attention of a variety of insects. As they come to search out their nourishment they may also carry pollen from a previous flower which they have recently visited. In this way pollination will be effected. Sadly, the life of the beautiful wild rose flower is all too short. Set there by nature as an attraction for the

RABBIT
The population explosion before the advent of myxomatosis meant that the rabbit spread into wild areas in towns. There are still many places—railway embankments, for example—where it can happily make its home. As long as there is a supply of grass and other herbs on which it can browse, it manages to do well. It has been estimated that each rabbit will need about half a kilogram of fresh green food every day.

insects, once pollination has been completed the job is over. The petals fade, and slowly over the next few months the familiar rose hip will develop. Although most of these have a distinctive oval outline, others present a more rounded form.

The outside flesh is tough, but once it has been broken the seeds contained inside will be liberated. Careful examination reveals that each seed is coated with a number of short hairs. It is still the tradition in some parts of the country to put the seeds down people's neck. They prove to be an effective form of itching powder!

Distinctive berries

The showy flowers of both the wild rose and the elder ensures that we notice them at other times of the year, as well as in the autumn; but this is not so with all plants. Some, like the woody nightshade, are virtually inconspicuous for much of the year. This plant, however, takes on a particularly attractive form in the autumn, with its red berries. Although the plant produces an abundance of berries, inside which is a good supply of seeds, nevertheless the woody night-shade will come up year after year in the same place, because of its underground stem. A climber, the weak

COMMON SHREW

In spite of its slight build, the common shrew can be aggressive and many people credit it with being one of the most quarrelsome of mammals for its size. Having a high metabolic rate, the shrew spends much of its life searching for food, and even in severe conditions it will be out and about after its next meal. Extreme cold results in the death of large numbers.

stem prevents the plant from supporting itself. It has to seek the support which other species offer it, or even the man-made supports like fences and railings, which might occur along the railway embankment.

Two distinctly-shaped types of leaf can be found on the woody nightshade. Some of these are shaped like the head of a spear; others are noticeably heart-shaped. The flowers are often overlooked. Purple and yellow in colour, they bear a resemblance to those of the cultivated potato. This is to be expected since the two species are related. Once fertilisation has taken place the fruits start to develop. At first they are green and shaped like miniature eggs. A variety of gradual colour changes occurs until they pass through orange to a beautiful almost coral red. The leaves die away and the berries hang in long strings against their supports.

As well as being known as woody nightshade, the plant is also called bittersweet. It is said that the berries have acquired this name because initially they have a bitter taste, which gradually becomes sweeter as they stay on the plant – but they should not be eaten by humans. Birds find the colour attractive, and come in search of the berries, helping to scatter the seeds.

Decaying agents

Other important agents are at work along the line, most of which will be active without the passer-by being aware of their presence. Timber occurs in varying amounts. Fencing posts and trees provide a certain amount. A variety of very small plants may plants, correctly termed fungi, need differing conditions depending on their species. Timber in the open is subjected to the weather, and unless it is treated starts to rot. It is generally easy to spot timbers which have been subjected to these fungal attacks. When they are wet they exhibit a dark brown or black colouration.

One of the most common of these fungi is a yellow species which is called *Coniophora cerebella*. For it to operate most effectively the timber on which it becomes established needs to have a water content of between 30 and 50 per cent.

Everyone knows about dry rot. This is caused by another fungus, *Merulius lacrymans*. It can work effectively in timber where the water content is as low as 20 per cent. Once it becomes established here it begins to attack dry timber as well. It manages to do this because, as water is released when timber is broken down, the fungus can make use of this.

As the dry rot fungus increases in size it can spread through cracks and crannies. In addition to this, it produces a vast number of spores. Once these find a suitable substrate they will begin to grow. Although softwoods are generally the domain of *Merulius lacrymans*, the fungus may attack some hardwoods such as oak and beech.

Once established, dry rot brings about a disintegration of timber fairly quickly. There are other fungi, however, which become established in timber. These work very slowly, and have generally come to be known as mildew and mould. Whereas some fungus has to penetrate into the cells of the timber to work effectively, other species live only on the outside of timber. Here they will survive in the very wet conditions on the external surface.

Other large species of fungus, including bracket fungus, become established on living and dead wood. Bracket fungus may bring about the death of a tree, and then continue to cause decay. On dead wood it is likely to help the tree to rot, and in doing, allows other species of a variety of fungal plants, as well as animals, to come in.

SPACE TO LET

Wildlife in parks and gardens

Even in the most densely populated urban areas there are some parks and open spaces. However small the areas, most of these enclosed spaces abound with wildlife.

By far the most easily noticed plants in the parks are trees. Where parks have trees, not only do they provide a habitat for wildlife, but they also serve another very useful function. No matter what their size, trees are valuable 'lungs' for towns and cities, giving out a much-needed supply of oxygen. Diseases, though, have taken their toll of urban trees, just as they have disappeared in the countryside at an alarming rate. Many areas have lost large numbers.

Trees are important to various species of wildlife. They provide food for a whole range of animals from the smaller invertebrates to larger mammals. Tree

LESSER SPOTTED WOODPECKER

The lesser spotted woodpecker is only slightly larger than a sparrow. Because of its size and secretive nature it is often difficult to discover. It is perhaps most conspicuous when young are in the nest, because they make loud calls.

BANDED SNAILS

Although banded snails seek the security of shaded places, they frequently come out. Banding and colouration varies considerably. Thrushes' anvils, around which there is often an assortment of banded snail shells, gives an indication of the predominant colour caught by the birds.

species found in parks vary from one place to another, and the value of each is different. A long-established 'native' tree, like the oak (*Quercus* spp.) has a whole host of invertebrate life associated with it, with as many as 300 different species using it for food – although not necessarily at the same time. In the case of the oak, mammals such as the squirrel will feed on the acorns.

The city oxygen-giver
In many built-up areas one tree is typical of the town environment: this is the plane (*Platanus x hispanica*), or to be more precise the London plane. This particular species is a hybrid between two trees – the oriental plane and the American plane. The resultant offspring is one which stands up well to the demanding and often dirty conditions in the town. As its name suggests, the tree is not a native species, the first being grown in this country in the seventeenth century. Some of the larger trees now encountered in towns may be up to two hundred years old.

The plane is particularly tolerant of the polluted atmosphere in towns. Bombarded with the impurities to which the London plane is subjected many other species would probably die. The most interesting feature of the tree is its ability to shed its bark. It is thought that along with the bark many of the unwanted impurities will also be shed. With its bark in disarray, the tree looks particularly bedraggled at certain times of the year.

A varying habitat for wildlife
Each park is different: some have been purposely laid out, originally as private gardens, whereas other areas have been public places for centuries past. Their value to other forms of wildlife – the birds, mammals and smaller animals which can live there – will depend on the conditions which occur, and what is on offer in the way of shelter, food and so on. Birds which visit the urban garden will usually frequent the park too. Here they will find shelter, food and usually a place to nest.

Whether or not many species of birds nest within the park will depend on the nesting sites available, and the requirements of each species. Where trees and bushes are a feature, they will provide a variety of places to nest and breed.

The birds of parks
Quite often we associate only a few birds with the town, overlooking many other species which are likely to be encountered. Two of those which are frequently noted, although they are not necessarily the most common, particularly in coastal resorts, are the sparrow (*Passer domesticus*) and the pigeon (*Columba livia*). The sparrow is mentioned in chapter 1. The pigeon is

indeed *the* city bird. It is an almost universal resident, and not only will it be found in parks, but it is equally at home in the street and parading around the squares. The bird has adapted well to city life. It has learnt that man will feed it, and in many places it will eat out of his hand. In a lot of instances it is visitors who feed the birds, and although they have to be tolerated they are not necessarily particularly welcome to the residents. In spite of its attraction – at least to the tourist – the common or feral pigeon also brings with it a number of problems. The mess which they make on buildings is undoubtedly the biggest headache facing the authorities. For this reason, large numbers are captured each year. Drastic measures in certain areas seem to have had little effect on the pigeon population. It would seem that the birds will increase and make use of as much space as is available to them.

PARKS – ECOLOGICAL PARKS TRUST

The Ecological Parks Trust has been established, and runs the first Park of its kind which is situated in Vine Lane, near HMS Belfast in the heart of dockland. Other projects are in hand, and the Trust advises various groups in other parts of the country.

For details of the Trust and its work, write (enclosing a stamped addressed envelope) to Ecological Parks Trust, c/o Linnean Society, Burlington House, Piccadilly, London W1V 0LQ.

The William Curtis Ecological Park in Vicar Lane is open seven days a week from 10 a.m. to 6 p.m. (sunset if earlier). The telephone number is 01-403 2078 for advice on finding the Park.

Pigeon problems elsewhere

It is not only the British who suffer from the pigeon problem. The birds have taken up residence in many American cities and Australian towns, and in other parts of the world. It would seem that they survive well with man. In Britain it is likely that they have come into towns as soon as they were built. In the case of London, we are not really sure when the feral pigeon made its debut, but records for 1285 show that it was already nesting on old St. Paul's, so it seems possible that it had already been in the city long before then.

A cave-dwelling ancestor

It is not surprising that the pigeon can survive in the brick and mortar jungle, for the bird's ancestors – the wild rock doves – made their homes in caves, rocky ledges providing suitable nesting sites. There are numerous similar ideal man-made positions in our towns.

Food is no problem for the city-dwelling pigeon. Able to adapt its diet to what is available, for much of the year, and particularly in winter, the bird depends on man's waste. At other times natural food will be found in parks, and large flocks will feed here. If there is a shortage of food, they may make their way to the outskirts of towns, where fields will provide an alternative source, and they can become pests, taking large quantities of grain.

An opportunist bird

Living in close proximity with man has turned many creatures into opportunists, as is the case with the wood pigeon (*Columba palumbus*). In open countryside the birds may seldom be seen, but in the town they have taken advantage of man. Park trees provide nesting sites for some birds: others use buildings.

One feature shared by few other birds is the wood pigeon's method of drinking. Dipping its bill into a pool of water, it can suck up the water and swallow it without tipping back its head, which most other species have to do.

A newcomer to the town

One relative newcomer is the collared dove (*Streptopelia decaocto*), which has now exploited the town environment. First breeding in the Norfolk coastal town of Cromer in 1955, it has now spread to much of the British Isles. Frequently found in towns, its identity can hardly be mistaken. Its striking collar – from which it gets its name – is its distinguishing feature. The white-fringed black band almost encircles the bird's neck. The plumage on the under surface is lighter than the grey-brown above. The tail is long, and the feathers underneath are black and white. Both cock and hen birds sport the same plumage, and so are indistinguishable.

1 ENGLISH ELM
The English elm grows to a height of 30m (96ft). It has been very badly hit by the ravages of Dutch elm disease.
2 WHITEBEAM
There are many varieties of cultivated whitebeam, which seldom reach heights greater than 10m (32ft). Old trees have smooth, reddish bark; in young specimens it is grey.
3 SYCAMORE
In the autumn the sycamore is easily identified because of its winged seeds, which aid dispersal.
4 OAK
Where oaks occur in towns they will provide small mammals including squirrels with a useful supply of food.
5 LIME
The common lime is a hybrid between two other species and is frequently encountered in many built-up areas.

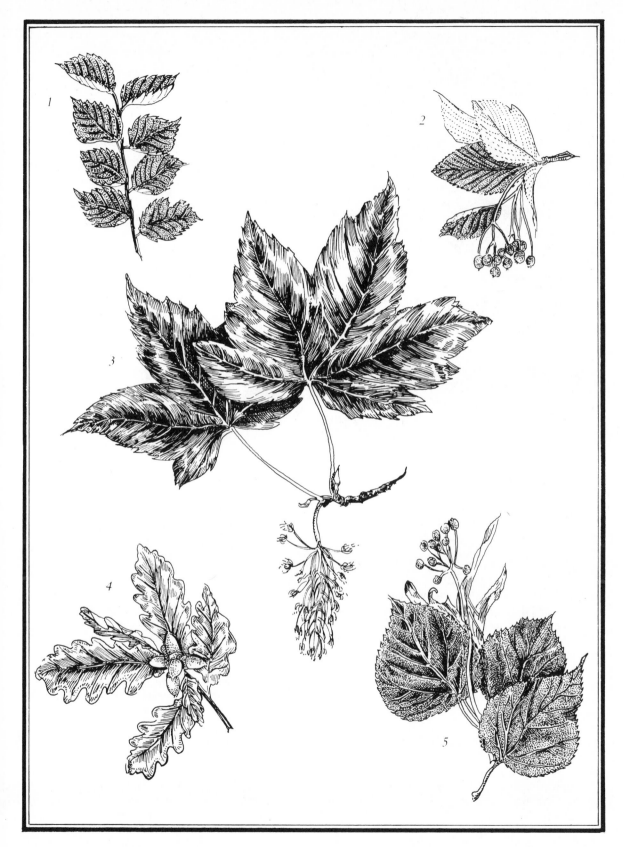

Although the collared dove has invaded almost every available space in our shores in twenty-five years, the reason for its remarkable success, when other species have failed, is not fully understood. Its ability to adapt and make the most of new environments has obviously a great deal to do with it. Man is a wasteful creature, and the dove has discovered this.

When it is time for nest-building, both birds assist with its construction, using a mixture of small twigs and the softer stems of other plants to form the home for the eggs and nestlings. The bird seems equally at home laying its eggs on buildings or in trees. In extreme cases, more novel situations have been chosen, and even telegraph poles have not escaped the notice of the bird in its search for a nest site. The female lays two white eggs, and the early birds will have theirs in the nest by March, whereas it may be September before others have achieved the same state.

Feeding chiefly on grain and the seeds of a variety of weeds, the birds will also take fruits, including elderberries.

Smaller species too

Apart from the larger species, many other birds will frequent the town. Whereas the inner cities will produce a shorter species list, the suburbs will boast many more varieties. A lot of these will come to the parks, and where wild gardens have been established, especially with a regular supply of food, the number of species increases. Some will visit the park by design; others by accident.

Where large expanses of grass feature in the city park, the mistle thrushes (*Turdus viscivorus*) will be out in search of food. Although they will be found at many different times, a shower of rain, which brings up the worms, will bring out the birds.

A charming species

Occurring in many urban habitats, the blue tit (*Parus caeruleus*) is equally at home in the park and the garden. A nervous bird, it needs the cover of small bushes and/or trees, so that it can move from one area to another as inconspicuously as possible. Although bold nervous and suspicious disposition, even to the extent of making off should the slightest noise or movement disturb it. Yet, with its almost cheeky character, it endears itself to most people.

Like many other British birds the blue tit was a bird of the woods, and it is quite usual to find them back in this habitat in spring. Many nest in the country, returning with their broods to the garden for easy living off the bird table. In the wood the bird has plenty of cover, and it is because of its change of habitat that it has this nervous manner, always seeking out a place in which to hide.

Given good days, the blue tit may start its melodic trilling as early as January. It needs a natural hole in a tree or a gap in a wall, but has taken readily to man-made nesting boxes to rear its young. Mates chosen, both cock and hen birds set to and prepare a suitable nest, first for the eggs and then for the young birds, using a mixture of material, including grass, moss, hair and wool. The female will lay in April or May. A small clutch will consist of around eight eggs; a large one almost double that number. As with many other hole nesters, the eggs are light in colour, the white surface being broken by a mixture of reddish-brown spots. Once the eggs are laid, the female takes it upon herself to be responsible for the incubatory duties. The male comes into his own once the eggs have hatched and the nestlings need feeding. He will join the female in searching for a continuous supply of food, and together in some respects, this bird does have a distinctly

STOCK DOVE (*Columba oenas*)
The urban status of the stock dove remains something of a mystery. Perhaps this is because to most people it is not unlike the feral pigeon, with which it feeds. It can be differentiated from its relative, because of the grey above its tail on the upper surface.

Old buildings provide it with a nesting site, and where there are holes in trees it may use these.

When W. H. Hudson wrote his *Birds of London* in 1898, he considered that the stock dove had only recently arrived in the city. Gradually its numbers increased here, and in other urban environments until late in the 1950s, when it started to decline.

During the day it often leaves its urban haunts in search of food—mainly grain—in the countryside. Sprays used on cereals probably took their toll of stock doves. Soon the population declined quite rapidly. The bird disappeared from many of its former nesting places. In the early 1970s, there was a sign of recovery, and this has continued.

they ensure that the nestlings will be kept supplied.

Mammals look for sites

The mammals which are found vary from one park to another, and one of the features which will determine their presence is the cover the plants provide. Of the larger mammals those found will fall into one of two groups. These are the introduced type – mainly deer – and those which have made their way into the park under their own steam as it were.

With smaller broods visits are made regularly. Sometimes the nestlings will not take the food, and the adults will eat it themselves. In large broods almost every visit is to satisfy the nestlings' needs.

Adult birds rely mainly on insects and other invertebrates for food, but they will take some fruits and seeds, as their antics at nut baskets testify.

The red and grey squirrels

Red (*Sciurus vulgaris*) and grey (*S. carolinensis*) squirrels fall into the latter group. Although there are places where the red can still be encountered, its general status is now one of decline while the grey in some places has reached plague or pest proportions. It is particularly sad that the numbers of red squirrels should have decreased so dramatically, since it is the native species, with the grey being brought in from the United States of America just over one hundred years ago. Although now a wild animal, when the squirrel was brought to this country it was introduced to be exhibited in zoological collections. Within a short time the first animals had escaped, and breeding colonies were soon established in many parts of the country.

Although the grey squirrel needs wooded countryside – mixed or broad-leaved woodland for preference – it has seized the opportunity offered by town trees. It is not unlikely that the first encounter a country dweller will have with a grey squirrel will be in the town, rather than in the rural environment.

Measuring around 45cm (18in) in length, of which 20cm (8in) is tail, at a distance the mammal's coat is grey, but a closer look at different times of the year shows a great deal of variation. Perhaps it would be more accurate to describe the coat as a mixture of grey and brown hair. The underparts are lighter: often almost a dirty white. It is in winter that the coat is more typically grey, and at this time of the year there is a border of white hairs to the tail. Although there may be hints of a reddish colour in the fur, the red and grey squirrels do not interbreed – as far as is known.

The fact that the grey squirrel can adapt to the often more difficult conditions in the urban set-up is probably due to its ability to vary its diet. Ideally it seems to prefer nuts, acorns and seeds, but the diet varies from

BLUE TIT
Few people will begrudge the blue tit its daily top of the bottle from the pint of milk on the doorstep. A regular visitor to the garden seed basket, the blue tit seldom flies in a straight line, but darts from one area of cover to another.

RAVEN (*Corvus corax*)

Having a diet which consists mainly of carrion, the raven will spend a great deal of time on the lookout for dead bodies, although they will take almost any kind of food which they can find.

A few years ago ravens nested in Cardiff, and this is a reminder that at one time it was in towns that they were frequently found. Here they discovered plenty to eat, especially in the streets of large cities.

Tame ravens are kept at the Tower of London to protect this land of ours. If they happen to leave, then Parliament will either fall, there will be chaos in the land, or our country will fall to a foreign power.

According to legend, the head of Brân—a name meaning raven—was brought from his native Wales, and taken to Tower Hill. It was duly buried there, its position such that it faced France. Safely beneath the ground, as long as it stayed buried it would prevent this island from being invaded.

is a debatable point whether they remember where their caches are. Nevertheless, they do not seem to go hungry, and each time they dig they manage to come up with something!

A typical rodent

The grey squirrel is a typical example of the rodent order. Rodents are gnawing animals, and the teeth have developed so that this operation can be effectively performed. The front incisors are elongated, and where they attack bark they can be responsible for a great deal of damage. By trying to get at the more succulent layers underneath they will often do untold damage to the outer covering. Where squirrels take off the bark all round the tree, leaving an open area, the tree will die.

The home in a tree is known as a drey – at least in the breeding season. In fact the animal has two kinds of home. In winter the grey squirrel may make large numbers of these homes, which are no more than stopping-off or resting points, as it moves about amongst the trees in search of food. The nursery drey, used during the breeding season, is a much more substantial structure, and is usually positioned wedged in the fork of a tree.

Each year there are two breeding seasons, with mating taking place between January and April, and from May to August. The average number of kittens in a litter is three, although births may vary from three to six, and two litters a year are not uncommon. At birth

season to season, and to it can be added bark, roots, shoots, fungi, bulbs and fruit, and from time to time even insects. In the town it tends to raid bird tables in gardens, and can become a nuisance. During times of excess food, the grey squirrel will bury it, although whether it does this as often in the town as country areas is questionable. According to many naturalists it the young are blind, and the ability to see comes after four or five weeks. Although many grey squirrels die early in life, the average life span is reckoned to be about twelve years.

Introduced deer

The other mammals likely to be encountered in town parks are the introduced species, the deer. Although brought by man, these mammals are now wild species. Fallow deer (*Dama dama*) are probably the most frequently found. Like all deer this species is adorned with antlers, but only the male bears these often large

ROBIN
Affectionately known as 'the gardener's friend', the robin has a more aggressive side to its seemingly placid nature. Once it has selected its territory, it will guard it fiercely. However, in times of hardship the robin is willing to leave in search of food.

and elaborate structures. Wandering around a park at certain times of the year, you may come across them lying on the ground, because they are shed once a year. During the part of the year when they are growing, they need a continuous supply of food, and this is carried to the antlers in the skin which covers them. It is this layer of skin which is known as velvet. Once the antlers are fully grown the skin has finished its job, and food is no longer required. Gradually at first the velvet flakes off. Once the antlers are fully grown the animal is ready to mate.

The colour of the animal's body is a reddish-brown, and the coat is marked with distinctive white, almost round spots. During the winter months, the spots tend to become more indistinct blotches, and these blend in well with the generally greyer coat.

The breeding season, known as the rut, takes place from the end of October through to the end of November. After mating the doe carries the young for around eight months. One or two fawns are born in the following June or July. The life expectation is between twenty and twenty-five years.

It is not quite certain when the fallow deer appeared in Britain. Some writers suggest it was brought by the Phoenicians; others seem convinced that it came with the Romans, or possibly later with the Norman invaders. No one is certain whether it was introduced to adorn the parks or for hunting purposes. Now, apart from park herds, there are groups living in almost every county in the British Isles.

Plants seek a foothold

The wild plants in many parks are discovered much less frequently than in many other habitats. Where grass is mown regularly this curtails the growth of these would-be rampant plants. Thus those which are likely to be found will have adapted to continual mowing. Many, like the daisy (*Bellis perennis*) have a rosette-type formation, in which the plants hug the ground. Others – the species present will depend to some extent on the seed mixture and initial colonisation – will be opportunists, still growing fairly close to the ground. In addition to these sorts of plants, there are others which will make the most of any opportunity to germinate and reach maturity. These will generally become established in those parts of parks where the mechanical mowers are unable to penetrate – around some trees, by fences and hedges, and so on.

Among the low-growing plants which are able to withstand constant, regular mowing, the daisy will be found quite often. Although very severe weather may curtail its activities, it is one of those species which manage to flower for almost the whole of the year. Indeed, its specific name, the perennis in *Bellis perennis*, gives an indication of this feature: *per* means 'through'; *annus* 'year'.

MUNTJAC (*Muntiacus reevesi*)

There are two species of muntjac deer – the Indian and the Chinese. Both species were kept at, and escaped from Woburn Park: the Indian species in 1890 and the Chinese in 1900. The former is probably no longer living in the wild.

Since then the animal has lived and bred in various parts of the country, and it is now encountered in urban situations, especially in London and the Home Counties.

It has the alternative name of barking deer because, when alarmed, it utters a sharp bark. At other times it growls. It is a small animal, measuring no more than 40cm (16in) at the shoulder.

Although there are probably large numbers of muntjac, especially in rural areas, it is an extremely secretive creature. This, coupled with its small size, makes it difficult to spot.

FALLOW DEER

The fallow deer, with its characteristic spotted coat, is found in many collections in built-up areas where it breeds successfully. A browser, for much of the year it will content itself with grass and other plants. In winter, when this may become scarce, it sometimes attacks the bark of young trees, and can cause considerable damage.

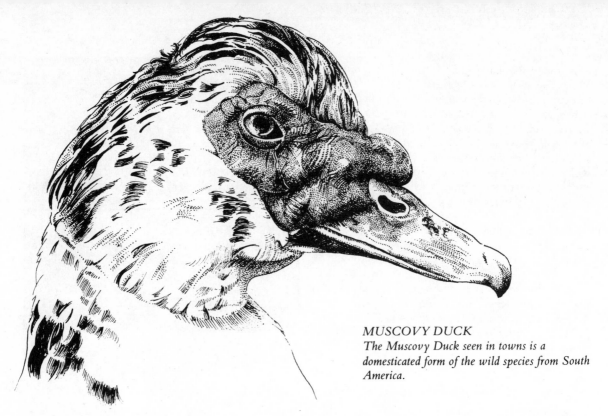

MUSCOVY DUCK
The Muscovy Duck seen in towns is a domesticated form of the wild species from South America.

Although the flower head appears at first sight to be a single bloom, it is made up of large numbers of florets. The flowers, sometimes brilliant white, sometimes tinged with pink, are borne aloft on strong stems. Where the grass is cut regularly the length of the stem will be short: where its growth is not restricted it may reach a height of 6–10cm (2–4in). Generally closing at night, its common name is derived from 'day's eye', a reference to the time when it was opened.

Straggler along the ground
Another plant which often manages to survive in among the grass is chickweed (*Stellaria media*). As with some other species, it is one of those plants which can usually be found flowering in the town for most, if not all, of the year. Although where there is some support, it may struggle upwards, in most places the weak stem causes the plant to straggle along the ground. When the small, white flowers are fertilised, they will produce seeds.

The 'universal' grasses
Grass is the most obvious feature of most parks. The variety of plants found will generally be determined by the original seed mixture which was sown. One grass which is included in most mixtures is annual meadow grass (*Poa annua*). Where it grows uninhibited, it will reach a height of 30cm (12in), but generally, where mowing is frequent, it seldom manages more than 5cm

(2in). Although, as its name implies, it is an annual grass, in some places it manages to survive through the winter.

Because of the almost inconspicuous nature of the flower heads of the grasses, most people miss them. Each flower is very small, and may appear on the grass at any time of the year. Because the flowers are pale in colour, they are not pollinated by insects, as are other brightly coloured species. Instead it is the wind which transfers the pollen from one flower to another. Even the slightest breeze will whisk these light pollen grains from one flower and with luck drop them on another. The pollen is the male sex cell, produced by the anthers.

A tough species
One of the toughest grasses to be found is the perennial rye grass (*Lolium perenne*). It flowers from May to August. The stem, almost straight, may grow up to 90cm (35in) where it is not cut. The leaves, usually referred to as blades, as in other grasses, are bright green, and it is quite usual for the under-surface to take on a distinct sheen.

Many cultivated species of rye have been produced: not only to improve the grass in parks and for garden lawns, but for agriculture. Rye grass is grown in many places for hay and in pastures where cattle graze. Strains have been produced which offer a greater yield than the original wild stock.

Frequently encountered around the edge of grassed

CAROLINA WOOD DUCK
A native from America, the Carolina wood duck may be found in some ornamental collections.

MANDARIN DUCK *(above)*
The mandarin duck is undoubtedly one of the most attractive waterfowl to be discovered by the town-dweller. The drake, with his brown, white, black and purple plumage, is striking. In contrast the duck is like that of other species—drab. Typically brown, she has a white ring around the eye, which is a significant distinguishing feature.

BLACK SWAN
An ornamental species, the black swan may be found in some collections on urban waters.

MAGPIE

Magpies, once country-dwellers, are now fairly common sights in many built-up areas. Attracted to objects which shine, magpies may make a large collection of these—much to the consternation of their owners.

areas, particularly where the mower is, fortunately, ineffective, is the distinctive cocksfoot grass (*Dactylis glomerata*). It is an easily recognisable species when in flower, because of the large inflorescence. It is from the shape of this that the grass gets its popular name.

Dwarf varieties

Many species of plants which grow to considerable heights where they are allowed to do so unhindered manage to survive in a different form in the park. Several species of thistle manage to grow in this way, although they are unlikely to flower and produce seeds. Some, like the creeping thistle (*Cirsium arvense*), increase by an underground creeping rootstock. Once thistles become established they are difficult to eradicate. With only a small part of the underground root still growing, they will produce a new plant. Male and female flowers occur on separate plants, so that cross-pollination is necessary.

GREY HERON (*Ardea cinerea*)

The only urban breeding colony of grey herons in Britain is found in Regent's Park in London. Although this heronry is small, the first birds moved in in 1968.

The grey species is the only kind of heron which breeds in the British Isles. In their search for food they may travel more than 20km (12 miles). In town and country they will go to garden ponds where they are not fussy what fish they take—including the owner's prize goldfish!

Once a site has been selected for a heronry, the birds will use it year after year. The nests are high in trees, and consist of a platform of sticks which the male collects, and the female arranges.

JAY

A colourful species, the jay is found where there are gardens and parks with plenty of tree cover. At the bird table, it takes almost anything which is offered. With the decrease in wooded areas the jay has had to find other haunts. It has gradually moved into built-up areas, and is now found well inside the boundaries of many towns and cities.

A typical plant of the park is black medick (*Medicago lupulina*). Belonging to the pea family, it becomes established in waste areas around the edge of the grass, but may also manage to survive in among the grass. As a trailing plant, with a spread of 5–50cm (2–20in) it manages to cover reasonable expanses.

Black medick and hop trefoil (*Trifolium campestre*), another member of the pea family, are often confused. It is possible to distinguish black medick when the seed capsules appear. They are black and have very distinctive veins. A relative of the clover, the black medick has a clover-like leaf divided into three. The whole plant has a liberal covering of hairs.

Fairy ring

Have you ever seen a fairy ring in your local park? A fairy ring consists of a conspicuous circle of darker green in the grass which can be seen particularly in autumn. Sometimes a ring of toadstools can be seen growing in the area. Fairies were supposed to dance around inside the ring at night, the toadstools providing them with seats when their activities have exhausted them. The further claim is that a lighter, almost bare circle of grass was due to dancing feet of the fairies as their constant movement wore away the grass.

If you do see a ring, and can keep details of it, you will find that it will get bigger each year! As the underground mycelium of the fungi grows, it spreads

outwards. Starting life in one spot as a single microscopic spore it gradually spreads out over the years to encompass larger areas. The bare patch shows where the mycelium is beneath the surface.

A number of species of fungi give rise to this phenomenon, but not the common mushroom (*Agaricus campestris*), although the latter may grow in a circle, because of the spread of the underground mycelium. You might be lucky enough to discover this species growing in your park.

In the town garden
Gardens, even among the greatest concentration of buildings, support many species of wildlife. Whereas plants will be either a permanent or semi-permanent feature, depending on the attitude and tolerance of the gardener, the animals may simply be visitors.

A butterfly may flit in and out, or perhaps stop momentarily in search of food: a bird may come and eat from the bird table. In the hedge, a blackbird may find a nesting place, or a snail shelter, and a wide variety of animals, from the large mammals to the small invertebrates, will probably find a supply of food on the plants, both wild and cultivated.

In many gardens, a natural system of biological balance can still be seen. Most gardens sport roses of some sort, and as we all know these beautiful cultivated flowers are subjected to the ravages of the aphid or greenfly (*Siphocoryne rosarum*); but generally it is not only the greenfly which will find a plentiful supply of food, especially during the warmer months of the year.

1 *WEEPING WILLOW*
All weeping willow trees found in the British Isles are female. Imported from China and south-west Asia, it is one of the more popular of the cultivated species.

2 *ALDER*
Often said to live with its 'feet in water', the alder will be found where there are moist conditions. Although not a conifer, it does produce cones in the autumn.

3 *HORSE CHESTNUT*
There is something distinctive about the horse chestnut at every season. In winter it is the sticky buds and in spring the flowers, affectionately referred to as 'candles' by many people. In autumn the tree becomes perhaps most popular when everyone seems to be out conker hunting.

4 *BEECH*
The beech is a native tree found throughout the British Isles. Liking a chalky soil, it thrives in parks and suburban areas. Its edible nuts make it most attractive to smaller mammals such as squirrels and wood mice.

5 *LONDON PLANE*
The London plane is a hybrid between the Oriental and American planes.

GREAT SPOTTED WOODPECKER
At one time it was thought that the great spotted woodpecker's days were numbered, since it became extinct in Scotland and northern parts of England at the beginning of the last century. However, it has made a spectacular comeback, and has now made its way into the centre of many cities. It appears to be adaptable, feeding on a wide range of food, and because of this aspect of its behaviour it has managed to expand its territory.

aphids, will do the work quite adequately and replace the poisonous spray.

The population of aphids is often controlled by the preceding winter's temperatures: when conditions are mild many will survive, but severe winters will drastically reduce if not totally eradicate the population. With large numbers surviving, the chances are that there will be a population explosion. A similar situation applies to the greenflies' controller, the ladybird. In mild winters large numbers survive, and even in severe conditions there are still some which manage to withstand the rigours of continually cold and wet weather.

Why are aphids so dreaded by the gardeners? They feed by sinking their piercing mouthparts into the stems of plants, withdrawing sap, which provides them with a source of food. It is also known that they carry diseases from one plant to another.

Those which survive the winter are joined by new insects in the spring. The adults will have laid eggs in the previous autumn, and when the temperature rises these will hatch out. From some of these eggs will emerge wingless females. Within a short time these too will be increasing the aphid population by producing new generations. These females give birth to live young. At this stage they do not need to mate, and so males are superfluous to their activities.

Some of the new insects will be winged females, and in this way the insects move to pastures new. Here they will soon be repeating the life cycle. It is not until much later in the season that males will appear when they are necessary to fertilise the females which will lay eggs for the following spring.

Ladybirds are not the only animals which find aphids useful. Ants are also attracted to the insect. They do not devour them as the ladybird does, but feed on the honey-dew – a sweet-tasting liquid – which the aphid produces.

The most familiar of the ladybirds, and the one seen most frequently in the garden, is undoubtedly the red seven-spot variety (*Coccinella 7-punctata*). As with the majority of species, the seven-spot is a carnivore, feeding, like many of its relatives, on aphids.

Many species of insects which feed in the open, and are likely to be attacked by predators, exhibit bright colours. The seven-spot ladybird is a typical example of an insect which has warning colouration. As an insect, the ladybird has the typical four stages in its life history – egg, larva, pupa and adult. Warning colours are shown in the larval stage as well as in the adult form, a kind of 'stay away because I'm distasteful' adaptation. Those ladybirds which are disturbed have a second form of defence. They are capable of producing a repugnant orange-coloured liquid.

TREECREEPER

It is often difficult to discover the whereabouts of the treecreeper. Although it is relatively common, especially in some urban parks, its brown plumage effectively camouflages it against the tree trunk on which it is working. When disturbed, it has a good turn of speed and can vanish from sight very quickly.

At these times, when conditions for reproduction are favourable, the vegetable gardener will find blackfly (*Aphis fabae*) on his beans.

In some instances the gardener will do all sorts of things to get rid of his pests; but some gardeners may not need to do a great deal, because they will have a natural controller. The ladybird, which feeds solely on

Putting the holes in leaves

Neat, smoothly rounded holes in the leaves of a variety of plants, including roses, usually signify the work of the female leafcutter bee. There are several species responsible for this activity, but the most common is *Megachile centuncularis*. Unlike honey bees, the leaf-cutters are solitary. The bees cut out pieces of leaves to line their burrows and for the partitioning off of each cell.

After mating the female finds a suitable burrow or, if she is unsuccessful, digs one for herself. Here she makes a number of cells in which she puts a mixture of pollen and honey, which provides food for the larvae once they hatch from the eggs. When they hatch the bees await their turn to emerge from their cells, leaving those in front to make their exit first. The cells closest to

GREEN WOODPECKER
The largest woodpecker in the British Isles, the green woodpecker finds habitats within the town which are suitable for breeding. Although many may breed on the fringes of built-up areas, others penetrate further inland to breed in parks.

the entrance always contain male bees; those towards the back of the burrow female ones.

Organised social insects

Communal living in the ant colony is much more effective than even the most efficient commune. Irrespective of their species, ants are hardly likely to be mistaken for other insects. Their distinctive feature is undoubtedly a body which seems to be divided into two with, in some species, but the tiniest strip of skin connecting the two parts. They also have antennae which are bent, and the head is large in comparison with the rest of the body and other insect species.

The ants have a social order, with a number of different members each with its own job to do. Unlike the wasps' colonies, theirs are more or less permanent; they do not disintegrate at the end of each season. As with wasps, the head (or heads in some colonies) of the household is a queen.

The queen commands a vast army of workers, which have usually developed from eggs she herself laid, although on some occasions a queen ant may actually come into an established colony, where she seems to co-exist happily with the queen which was already 'in charge'.

There are no individual cells for the eggs when they are produced. Instead, after the queen has laid them the workers collect them together and place them in one particular area of the nest, usually closer to the surface, so that the sun's warmth will reach them. The ants will lick each one, which prevents fungus from attacking the eggs.

How long the eggs take to develop varies. In some instances the small, almost insignificant larvae, their bodies a pale yellow colour, may emerge in a few days; on other occasions the development time may run into weeks. The workers, ever eager to complete their chores, will spot the larvae as they are born. They are quickly transported from the egg-hatching area to a 'nursery'.

Apart from tending the eggs, the workers now have to feed the young. They collect together food, some of which they will swallow and then regurgitate. As the larvae grow, the demand for food increases. Eventually they will be fully grown and ready to pupate. It is this stage which is often falsely referred to as 'ant-eggs'. Perhaps such a mistake is not unreasonable, because the pupae, wrapped in a cocoon for protection, look like large white eggs. Each pupa will generally spend between fourteen and twenty-one days encased in its cocoon. When they are ready to emerge, the ever-attentive workers are ready, and they will be on hand to break open the 'shell'. At certain times of the year cocoons larger than the majority appear, and these will contain the pupa of a queen. Although there is nothing to distinguish them from those of the workers, some of the smaller cocoons will protect winged males. These are only found when the queens are in the pupal state.

When it is time for the mating flight, the queen will leave the nest, and may climb up a suitable plant stem, as if a launching pad is necessary for her to take off. She and her pursuers will attempt to take to the wing. Her exit from the nest is not without its difficulties. Although she is eager to leave, both she and the males are often prevented from doing so by the workers. It is these which decide when environmental conditions

outside the nest are at their most favourable: conditions which will be most advantageous for mating. Similar restraints will be imposed on queens in other nests. Thus each ants' nest in one area will be ready to release its queens and males at the same time. This gives rise to the familiar 'plagues' of flying ants. Rid of the queen and the males, it will now be necessary for the colonies from which the ants have left to get back to normal.

Once she has mated the queen's power of flight is no longer necessary. She is fertile and ready either to start a new colony or to join an existing one. She has no need for her wings, and will get rid of them. She finds a

LITTLE OWL
Now a well-established species, the little owl has only been part of the fauna of the British Isles for the last hundred years or so. It was brought to our shores in order to act as a natural pest controller, with the hope that it would take small mammals and insects. Although nocturnal, the little owl is abroad during the hours of daylight—far more often than any other owl species. The bird tends to stay near the edges of towns, presumably in order to make its way outwards when feeding. It nests in large parks.

suitable object and brushes the wings against this until they fall off.

Now she seeks a place where she can remain in solitude for several months. It is necessary for the eggs to mature before they are laid. During her solitary state she does not need actively to seek food. Now that she has no wings, she does not need her large flight muscles. Whilst she is relatively inactive there is enough energy in these to provide her with her food.

As spring approaches, her time for egg-laying draws near. Her first young will all be worker ants, which she manages to feed when they are first born on her own saliva. This seems nourishing enough, and the larvae ultimately pupate to give forth the first worker ants for the new colony.

The first and most important task which these newly-emerged workers have to undertake is the construction of a nest for themselves and their leader. As far as most of the ants found in the garden are concerned, their nests will be under stones or under a path, the majority, if not all, of the nest being constructed below the surface.

In spite of the ants' industry, the nest which these workers construct is much less elaborate than that which either social bees or wasps produce. In the soil

they will burrow out many chambers which are linked by innumerable corridors. Any soil which is pushed up as a result of these underground activities can easily be seen. It is not unusual for the chambers and tunnels to continue into these heaps of fine soil.

The ants are highly organised, each chamber has its function and the ants know their jobs, going about them methodically and efficiently. Because some of the chambers which house the eggs, larvae and pupae are close to the surface, they are often accidentally attacked by the gardener, but the ants seem to cope with what, to such small insects, must be a major disaster. Within a short time they have collected together the incapacitated occupants, often scattered over a wide

distance, and placed them back in their chambers. Whilst one group of workers completes this, another band repairs the damage to the nest.

How the ants know what their particular job is remains something of a mystery. Entomologists who have watched ant activity are convinced that they have a means of communicating. Observation of ants will soon reveal that on their many wanderings they stop frequently as they pass each other. Antennae are rubbed together and this seems to be a means of passing on information.

Nature's protective function

Mimicry is one of nature's ways of ensuring that otherwise unprotected species will survive. In general mimicry takes the form of bright colours like those exhibited by other known distasteful species. Of those found in the garden, perhaps the most distinctive group is that of the hoverflies. Boldly marked with yellow and black, at first sight they are often mistaken for wasps. Nature has decreed that by adopting such a strategy they will be afforded some protection from would-be predators. In the early stages of life it seems that birds, for example, would endeavour to eat all kinds of insects. Within a short while they have found those which were palatable and those which were not. Thus, finding that wasps are not good food, they must identify all similarly coloured insects with an unpleasant experience, and so leave them alone.

Hoverflies are aptly named, and can be distinguished from wasps by their ability to hover. Indeed they are able to hover in one place for several minutes should the need arise. All hovering insects are grouped together under the heading 'hoverflies', but there are many different species, although they are generally identified by their Latin names only. One of those frequently encountered in gardens and parks is *Syrphus ribesii*. It visits flowers to extract the nectar which it needs for food, but it will often rest awhile on the leaves, especially on a sunny day. Although the larvae of many hoverflies are useful friends of the gardener, feeding on the destructive aphids, there are others which are not so welcome. One in particular has larvae which take their nourishment from daffodil bulbs under the soil. In particularly bad attacks, the larvae may take so much of the bulb that it dies off. This species is appropriately known as the narcissus fly (*Merodon equestris*).

Butterflies

Butterflies of many species will probably visit the garden. Whether they stay will depend to some extent on whether there is food for either themselves or their caterpillars. The common butterflies – small tortoiseshell, peacock, red admiral, wall brown – apart from coming to lay their eggs, will also come in search of food. Although they may flit in and out from spring through to autumn, they often occur in the greatest numbers in late summer and early autumn. At this time of the year it is the michaelmas daisies and the buddleia which provide a good supply of nectar. Those species of butterfly which over-winter as adults will need a good supply of food at this time of the year to see them through the difficult months ahead.

Provided that there is some form of protective shelter – such as an outhouse, garage or cellar – some species may over-winter somewhere in the vicinity. In spring, watch out for the species which emerge early.

Fruit lovers

There are other animals which, although present for much of the year, become particularly conspicuous at certain seasons. Of these, sawflies (Cimbicidae) will probably be seen from time to time, the larvae being found more often than the adults. These larvae are often mistaken for caterpillars, but closer investigation will show that they have more legs towards the rear end of the body. Most species are herbivorous, feeding on a variety of plant leaves. There are others which actually bore their way into plants – both leaves and stems. When they do this they will often cause galls to appear. It is because of the irritation which the sawfly larva sets up that the plant will produce a gall. This swelling will provide the nourishment which the sawfly larva needs.

Cultivated fruit seems to suffer in some areas. Both apples and gooseberries have sawflies associated with them. Whereas the apple sawfly buries its way into the leaves of the fruit tree, the gooseberry sawfly larvae work by devouring large numbers of leaves of both this fruit bush and black and red currants. Whole branches, and in extreme cases complete bushes, may be stripped of their leaves.

The female sawfly is equipped with an effective egg-laying device, called on ovipositor. In most these resemble a miniature saw, giving the insect its name, and using this she easily opens up plant leaves and stems so that there is a place to lay her eggs.

A nocturnal creature

Have you ever wondered how those common garden creatures, the earwigs, got their name? It has been suggested that the insect was originally called an ear wing. This was due to the ear-like shape of the hind wings. The other popular explanation is that earwigs sometimes managed to crawl into people's ears, especially when they slept in the open.

Although the earwig is a particularly common garden creature, few people bother to take a look at this intriguing insect. Although the species which is found most often is, as would be expected, the common earwig (*Forficula auricularia*), there are several other species to be found in the British Isles.

A nocturnal creature, the earwig will often be found during the daytime when its hiding places are disturbed. Bricks, flower pots, sacking, wood – in fact almost anything in the garden – may conceal one or more earwigs. When the earwig comes out to feed, it will devour virtually anything it comes across. They take both living and dead food, sometimes attacking the petals of flowers.

Both male and female earwigs possess the well-known pincers. Not particularly easy to distinguish apart, those of the male are generally more curved than the female's. It has been suggested that the pincers are used to fold the wings, but entomologists generally agree that these are in fact defence weapons. Earwigs take on a menacing attitude if they are disturbed. In this case, the tail end is pushed up into the air, and the menacing pincers held wide apart – this appears to be more a threatening display than one of attack. If the animal accidentally falls onto its back then it may use its pincers to help right itself.

Pairing of male and female earwigs occurs towards the end of summer. Pregnant females will retain their eggs until they take their winter's rest. Sometimes the female will find the male tagging along with her. Once their eggs are laid, unlike most insects, the earwigs take great care of them. Tending them, the female has been observed licking the eggs from time to time. It has been suggested that by doing this she prevents bacteria from attacking the eggs. One particular observer moved the eggs on a number of occasions, and each time the earwig collected them together again with meticulous care.

The eggs will finally hatch out in the following spring, and the female has been seen to lick the young insects when they emerge. For the first two or three weeks the female will ensure that they have enough food, and once they are old enough they will join her on her food hunting expeditions. The young earwigs remain with the female until June. Apart from being smaller than her, the young also have a paler skin.

The 'little ploughmen'

Of all the animals to be found in the garden, there is one the work and importance of which should be appreciated by the gardener. This is the lowly earthworm. The great nineteenth-century naturalist Charles Darwin dubbed worms 'nature's little ploughmen', and this is what they are. Living in burrows

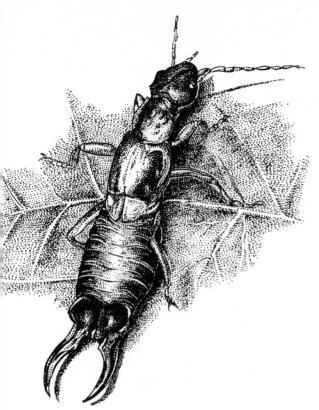

EARWIG
Males have curved pincers and females straight ones. The earwig uses them as defensive weapons, lifting them when danger threatens. The common earwig has wings, but seldom takes to the air. Some people suggest that this is because the wings are so large that it takes the earwig too long to fold them after flight.

show a small hole. It is through this that the male reproductive cells will pass as mating takes place. One segment in front of this are the female organs: female reproductive cells will come out from these.

A distinct swelling of the worm's body can be seen between segments 32 and 37: this is the area called the saddle. It is in the saddle region that the cocoon will be produced, and this will eventually enclose the eggs.

Mating usually takes place at night on the surface of the soil. During certain times of the year – particularly in autumn – when there are damp mornings, mating earthworms can sometimes be seen on the surface of the soil. Although worms have both male and female reproductive organs, they do need to pair off. Only a transfer of cells from one to the other will ensure the production of eggs and the ultimate emergence of young worms.

When they pair, the head end of one earthworm will be pointing towards the tail end of the other. Before they mate both earthworms produce a great deal of mucus which covers their bodies. In this position male reproductive cells pass from one worm to the other and *vice versa*. This complete they come apart. The saddles produce a membrane and when this is complete it becomes free. After eggs have entered it, it moves towards the head end as the earthworm wriggles backwards. As the cocoon moves in this direction sperms are added. The membrane passes over the head end, and after it leaves the earthworm's body the ends come together to form a sealed case, the cocoon. Within the cocoon it is normal for only one of the eggs to develop into a worm.

The earthworm is particularly adapted for movement within the soil, its under-surface having a double row of bristles which can be pushed in and out of the body. The earthworm moves by relaxing and contracting its muscles. With the bristles (*setae*) at the front end of the body anchored to the soil, it then pulls up the rear end. Now by gripping the sides of its tunnel with these *setae*, it can push out the front end of the body.

The earthworm breathes through its skin, and this is why the surface has to stay moist. If it dries out, the worm will not be able to breathe, and it will die. In the soil, where the worm lives for the majority of its life, it will find a moist environment.

Moving through the soil prevents no difficulty for the earthworm. Where the soil particles are compacted, the earthworm eats its way through. Indeed, this is one of the valuable services which the animal performs for the gardener. It also helps by pulling down the leaves from the surface. After softening the leaves, it will eat them and in this way, helps break them down and pass back nutrients into the soil in its waste.

beneath the soil, the earthworm has the ability to tunnel almost continuously. Throughout their lives they are actively ploughing up the soil and increasing its fertility.

Whilst digging about in the garden, one is likely to uncover many worms. Apart from the earthworm (*Lumbricus terrestris*) there are more than twenty other species called worms. At present we are concerned only with the earthworm.

Most will measure between 10cm (4in) and 25cm (10in). The head end is pointed and the tail end bulbous or flattened. A segmented invertebrate, the earthworm has around 150 divisions on its body. As would be expected, at the front end there is a mouth, at the rear end the anus.

Each earthworm carries both male and female reproductive organs: this makes them hermaphroditic. Fifteen segments from the mouth end you will see a paler colour and some swellings. Close investigations

SWALLOW

Its distinctive red chin sets the swallow apart from the swift and martins. It is not unusual for a swallow to return to the same nest for several years. Because flying insects, on which the bird feeds, change their habits according to the weather, the swallow must do the same. When insects fly high, so does the swallow; when these invertebrates fly low, the swallow follows.

SWIFT

Swifts perform superb aerobatics. They spend most of their life on the wing, where they are true masters of the air. In fact, with the exception of the need for incubation, they seldom return to land, even sleeping on the wing.

The tree – a home for all-comers

Many of the wild things which are constantly around us are so often overlooked. For some creatures, such as the earthworm which lives and work beneath the soil, this is understandable; but birds may build their nests, and go about their egg-laying and rearing of their young without anyone suspecting their presence; and so it goes on. Trees, however, are such large structures, and a dominating feature of the garden and park, that they cannot be overlooked. Or perhaps they can! We see them, but do we realise just how important they are for wildlife?

Established in a few older gardens, and in parks, is *the* British tree, the oak. Steadfast, sturdy and strong, the oak has matured from a tiny acorn into a massive dominating feature of the landscape. The creatures which seek out the oak do so for a number of reasons. Some come simply for the shelter which the tree provides; others come to feed. The oak has plenty to offer. In winter and early spring the buds are sought out by some creatures. In late spring and early summer the leaves provide nourishment for those seeking them. Later the fruits – acorns – provide a veritable treasure chest for many creatures. Out of sight beneath the surface of the soil the roots carry out their important functions unseen and unsung. Yet even these will be the target for certain species which find the roots good to eat. Bark and timber are not exempted.

The oak provides a home for no fewer than four hundred different invertebrates (animals without backbones). Of course it is highly unlikely that all of these will be found on one oak tree! Nevertheless, large numbers may occur, with the passer-by totally oblivious of the importance of the tree for these creatures.

To look at them all would in itself provide enough material for a book! To look at some, however, gives a fascinating insight into how one species of tree is important for wildlife.

By beginning at the bottom, and working our way upwards, we will gain some idea of life in the oak. Not far below the surface of the soil, some of the roots of the oak can be discovered. The oak is so large that in order to provide adequate support and anchorage the roots have to spread out a long way.

The root 'attackers'

The cockchafer eats the leaves of oak and the larvae of the insect attack the roots. Living below the soil, the larva of the cockchafer can be a serious nuisance, especially where there are young oak trees. With more established forms, it is likely that little damage will be done to the roots. As the tree gets older, the roots get tougher, and the larvae can only eat the thinner, younger, more succulent roots.

As the cockchafer grub continues to feed below the surface of the soil it grows into a fat grub. It has a white body and a brown head. In some places it is called the rookworm, simply because where they can find them rooks are partial to the larvae. Each larva may take between three and four years to grow fully, so that it is ready to pupate. This explains why there may be large numbers of cockchafer beetles every three or four years – it has taken this long for the larvae to mature and change into adults.

All the British species of 'swift' moths have one thing in common. The caterpillars feed below the surface on roots. The common swift species may sometimes eat the roots of oak. It is only the younger roots which are attacked. These are more nutritious and easy to bite through.

The two species just mentioned may find other roots on which to feed. There are other creatures, however, which can only survive as long as they have oak roots on which to live. These are gall wasps. Two species on oak roots are *Andricus quercus-radicis* and *Biorhiza pallida*. Below the surface of the soil the galls can be seen as irregular lumps on the roots. By removing one of these and cutting it open, it is possible to see whether anything is inside. Although it might be something else, it is usually safe to assume that if an egg, grub or chrysalis is inside it belongs to a cynipid gall wasp. The gall wasp will emerge as an adult after they have been inside the gall for two winters. It is possible to sort out the 'disused' galls from ones which are occupied, because there is a hole where the gall wasp has emerged. This may not always be a safe guide, since other creatures, realising that there is a free supply of food, may enter.

The galls of *Andricus quercus-radicis* are often called truffle galls. They are fairly easy to see since they measure between 3 and 6 centimetres across. These develop on the roots of young trees.

Defoliators

In spring and summer the leaves provide an abundance of food for a variety of creatures. The way in which the animals feed and the damage which they cause varies from species to species.

To most of us, aphids (greenflies and blackflies) are those insects which attack our roses and beans; but the oak suffers too. An oak tree may be infested with greenfly in the summer, although we seldom see them, because the lowest leaves may be too high above the ground. One species which commonly occurs on oak leaves is *Phylloxera quercus*. This and other aphid species feed because they have a sharp 'beak'. This is used to pierce the plant tissue, so that nutritious sap can be sucked up. In effect, although the aphids do obtain their nourishment from such feeding, they need to consume relatively vast quantities of liquid. There is also another problem. These aphids may carry disease which, when they pierce the tissue, is passed into the oak on which they are feeding.

Although the defoliators may escape detection, evidence of their activities is plain to see. In most areas there is something which attacks the oak leaves. Sometimes the onslaught is so great that the oak tree may be completely stripped of its leaves. If this happens early enough in the year the oak may produce new foliage.

In some places and in certain seasons the cockchafer may do untold damage to the oak tree. Adults fly in early summer, and have come to be known as may bugs for this reason. In fact the cockchafer is a beetle and not a bug. As dusk descends on the park or garden, the cockchafer takes to the wing. In the country the beetle seeks out solitary oaks; in the town they have it made. The activities of a large swarm can be so devastating that all the leaves on a very large oak may disappear in less than twenty-four hours!

Generally it is not the may bugs which do the most damage to the oak tree, but species of moth caterpillars. One which can ensure the complete loss of an oak tree's crop of leaves is the oak roller moth (*Tortrix viridana*). This common name is a fitting reference to the fact that the moth makes itself a home inside a leaf. Using silk it pulls together the edges of a leaf to form a tube. This provides it with suitable shelter. The other name for the caterpillar is green tortrix.

In a 'good' season almost all the leaves on an oak tree may be attacked in this way. Because it usually happens early enough in the season, however, the oak produces a second set of leaves.

If the caterpillars are disturbed once closeted inside their rolled-up leaf, they may leave the shelter of the tube, and hang suspended on a silken thread. Ready to pupate, the caterpillars go to the ground. It is an incredible sight to see thousands of capterpillars suspended from identical silken threads as they make their way to the ground.

Although this is perhaps the most noticeable of the caterpillars found on the oak tree, it is by no means the only one. Others will accept the tree's hospitality as fresh green succulent leaves invite them to feed. Many looper caterpillars can be seen on the tree. They have two sets of legs, one positioned near the front of the body and the other near the back, with a large gap between the two! To move, the loopers position the front legs and then draw up the back end, hence the looping effect. Many loopers are so well camouflaged that they are often mistaken for twigs. The correct name for this group of looper caterpillars is geometers.

NUTHATCH
Once resident in London's parks, in the nineteenth century the nuthatch left. The reason for its disappearance was a mystery at the time, although naturalists now suggest that it was probably due to increasing pollution. The bird returned to some of its former haunts in 1958. With its sharp, narrow bill, the nuthatch is able to penetrate cracks in between pieces of bark, and pull out insects lurking there.

This means 'earth measurers', describing the way in which they stretch out their bodies.

Several species of caterpillar may be encountered on oak. These include those of the winter moth, blotched emerald, common emerald, false mocha and mottled umber. When the latter species occurs in large numbers it can defoliate an oak as quickly as the oak roller.

Attacks on buds
The oak tree often displays a strange variety of growths called galls. They are caused by creatures, but made by the tree. Generally they are isolated, and there is no apparent damage to any other part of the tree.

Many of the gall causers on the oak belong to a group of very small insects called cynipid wasps. In fact the oak tree is subjected to attacks by more species of cynipid wasps than any other tree species. Not only do these creatures bring about a strange reaction when they attack the oak tree, but many often have unusual life histories. In the case of one species there are two different kinds of adults in the cycle. In one part of the cycle only females are produced. Yet, in spite of not having been fertilised, the eggs which are laid will hatch and give rise to both male and female forms. By far the most easily noticeable of the galls on the oak tree is the one referred to as the oak apple. It is not really an apple and neither is it the fruit of the oak, which is the acorn.

When the female of this gall wasp (*Biorhiza pallida*) emerges from the roots towards the end of winter, she makes her way to the oak bud where she lays her eggs. The egg hatches and the larva causes the bud to produce a soft, spongy material. It resembles a miniature apple, with pink and green colouring on the outside. This tissue provides the larva inside with plenty of food. By June, like many millions of other *Biorhiza pallida* larvae, it will be ready to emerge as an adult. Their first task is to escape from their gall-home, and biting mouthparts enable them to do this. There are both male and female cynipid wasps from this generation. There may also be many surprise visitors to the gall. Because the gall tissue had a good supply of food, which other species recognise, some may have moved in to take advantage of this free meal. The larvae themselves also provide food for certain predatory species, and their 'enemies' might enter for this reason.

Now that there is a male and female form, the first activity is to pair off and mate. Fertilised, the females are now ready to ensure that the second stage of the cycle will take place. They leave their mating places, and make their way to the soil. At the base of the oak tree they burrow into the ground, and lay their eggs on the oak tree's roots. As the larvae hatch the root produces galls. These provide food and shelter. Having spent two winters feeding here, the gall wasps are ready to emerge. They are all females, and make their way to the buds of the oak, so that the cycle can be repeated again.

The bark and timber of the oak tree provides another valuable source of food for some species. Just beneath the surface of the bark growth continues for some of the year. During spring and summer there is a great deal of activity here. The tree leaves need a constant supply of water, so that food can be made. This water travels up tubes just under the bark. Once the food has been made in the leaves it travels down the tree trunk. Here is a valuable source of food for some animals. Several species of bark beetles will make their homes just below the surface of the bark. The oak bark beetle (*Scolytus intricatus*) is related to the bark beetle

responsible for carrying the fungus which has caused havoc to our elm trees.

There are other species of bark beetles, all of which can be recognised by the shape of the chambers which they make beneath the surface of the bark. When male and female oak bark beetles are ready to mate, a pair lands on a suitable part of the oak tree. The female uses her strong jaws to make a hole in the bark, and into the part of the tree where growth is taking place. Both beetles enter. Once inside the males makes sure that a mating or nuptial chamber is excavated. Here mating takes place, and the female then makes a tunnel in which she lays her eggs. Once the eggs hatch the larvae are on their own. They make their own tunnels, eating the wood which contains a valuable supply of food. Once it has reached the fully grown stage, the larva needs somewhere to pupate. It enlarges the end of the tunnel in which it is feeding. Here it makes the transformation from larva through pupa to adult.

Holes for free

As oaks age they become even more important for wildlife. Holes in oak trees, especially in the urban situation, provide an attractive home. Several species of owl (all hole-nesters) may seek out the oak for their lay-laying. By far the most frequently encountered of those is the barn owl (*see page 133*).

Other birds also need holes for nesting sites. These include tree sparrows and starlings. Although not in need of a hole, the tree creeper may find a suitable niche behind the bark of the oak tree.

It is easy to forget that there might be temporary homes for some creatures. Hollows in oaks (and other trees as well) will collect rain drops, providing a small stagnant area of water. This usually contains material which has collected there for some time. In summer certain species of insects are desperate for any site which they might use for egg-laying. Even in this seemingly undesirable haunt it is possible to discover more than one species of gnat larvae, not to mention those of hoverflies.

In other sites on the wood the repeated showers of rain make some areas very damp. This provides a niche for such species as the larvae of the muscid-fly, craneflies (daddy long-legs) and the stiletto fly.

In autumn the acorns attract a variety of species. Squirrels take their fill, and the badger may stop to sample some on his nightly jaunts. Birds have their share. The jay is partial to acorns, as well as the wood pigeon and pheasant, and perhaps occasionally the woodpecker may decide to vary its diet.

Next time you pass a tree – and not just an oak – pause a while and try to imagine how important it is for a whole range of creatures.

BOUND TO PLEASE

Walls and hedges

Walls are usually ignored as wildlife habitats, but in sprawling built-up areas they provide a refuge for a wide spectrum of plant and animal life, even though the variety is restricted. The most interesting walls are, perhaps not surprisingly, the ones which are found in the older parts of towns, and this is particularly so where they are made from stone. Newer walls also provide important habitats. Even the walls of relatively new houses are subject to the interest of masonry bees!

Old walls provide a niche

In most areas by far the most prolific colonisers of walls are the lichens. Perhaps 'prolific' may not be quite the right word, because although they are found on many walls, their growth rate is extremely slow. Many species of lichen are unable to grow in polluted conditions, but with the creation of smokeless zones lichens have been able to establish themselves in new areas in towns and cities. At first sight they are not like the other more familiar plants; they bear little resemblance to species like grass and roses which we know so well. Yet in spite of their apparently lowly nature, they have a rather complicated life history. To start with they consist not of one but two types of plant, fungi and algae.

Although they may only grow a few millimetres each year, they are exceptionally long-lived. A large patch of lichen, particularly on a church, may be a thousand or so years old. Although lichens reproduce by means of very small spores, they also spread by

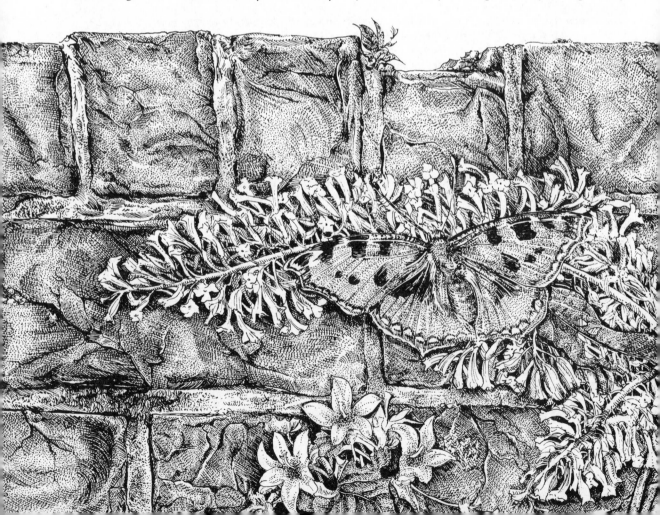

other means: small pieces break off, finger-like projection (called *isidia*) spread out, and granules, with a powdery texture (*soredia*), also ensure dispersal.

Colourful patches

One of the most conspicuous species, solely because of its colour, is *Candelariella vitellina* – some call it the mustard lichen. This, although common, usually only grows in small patches.

In southern parts of Britain a very common lichen *Buellia canescens* – the frosted wall lichen – is found.

This species tends to become rarer the further north one goes, and it is particularly infrequent in Scotland. *Lecidea lucida* (the sulphur lichen) is also common and does particularly well in cleaner towns where there is a moist situation. Thus it is found growing along the mortar between building blocks, and also in the inscriptions on gravestones, because there tends to be more moisture there.

Considered by many to be the most common species in Britain, *Lecanora dispersa* does particularly well on limestone, but will also be found on concrete, cement and mortar, as well as on asbestos roofs. Greyish-white in appearance, it is a crustose lichen. Two species which give some indication of the pollution – or rather the

THE WALL
Although generally far less hospitable than many other habitats, nevertheless the wall provides many species with a home. Seeds dropped by birds will begin to germinate, and these may include shrubs such as buddleia as well as smaller herbs such as slender thistle. Butterflies alight on the wall, because on a warm sunny day it acts like a radiatory. Here the small tortoiseshell, painted lady and comma seek to sun themselves.

lack of it – in towns, are the bright orange specimens *Xanthoria aureola* and *X. parietina*. The former is found particularly in eastern counties of Britain, but is much less common the further one moves west. A very deep orange, the plant was once used as a cure for jaundice. This species was also collected before modern chemical dyes were produced and it yielded a brown or yellow colour, used in the dyeing of cloth and wool. Natural craftsmen, especially spinners of wool, still collect this (and other plant species) as the basis for their dyes. *Xanthoria parietina* varies in colour from a bright orange to a more greenish-yellow colour, where it grows in the shade.

Mistaken for mosses

Damp walls provide a suitable environment for liverworts, and many occur at the junction between wall and road or pavement. Although several species may occur in the urban environment, by far the most common are the crescent cup liverwort (*Lunularia cruciata*) and the common liverwort (*Marchantia polymorpha*). At first sight it is not surprising if they are mistaken for the mosses to which they are related. They have fine, almost thread-like leaves, but careful closer examination will show that, unlike those of moss, they do not have a midrib. This fact ascertained, one can go on to look more closely at the leaves. If the leaf under scrutiny has two roughly equal lobes and v-shaped indentations (notches) along the edge, then there is little doubt that it is a liverwort.

Because of the make-up of the plant, these species are generally referred to as 'leafy liverworts'. There is another group, called thalloid liverworts, in which a distinctive flat disc, called a thallus, can be seen. This grows very close to the surface of the soil, and a number of hairs growing out from the under-surface afford it anchorage. Although these are not roots, they perform a similar function. However, since they are not root-like in their structure, they are called rhizoids.

The crescent cup liverwort is not a native species. It seems likely that it was transported to this country from the Mediterranean area several hundred years ago. It produces a number of small cup-like structures, which are not fully formed. Within these are the organs of reproduction (*gemmae*), which, once they leave the parent plant, especially when rain dislodges them, will start to grow as independent plants in adjacent areas.

Mosses form flat cushions

Most walls will support mosses. At first sight they appear uninteresting, but if viewed through a hand lens, they have an intricacy of shape and form which is a tribute to nature. Because more than one species may become established in one area, there are not always clearly defined plants, and this can make identification something of a problem.

As far as walls are concerned the number of species which is able to gain a foothold is limited and so this decreases the problems of identification. As one of the first plants to be able to live on the wall, moss starts the decay process. It can retain water, making the surface wet, which aids the break-up of the wall. As they do this they make more materials available for the larger plants which come in, establish themselves and survive.

In urban situations, the wall screw moss (*Tortula muralis*) will be found occurring on the tops of walls. Here it forms small tufts which may measure up to 3cm (1.2in) across. If the wall becomes dry, the moss will dry up and become twisted. Now the moss takes on a grey, rather than the normal green, colour.

Where cracks are available the silvery thread moss (*Bryum argenteum*) will find a suitable niche. The name

---

BORAGE (*Borago officinalis*)
At one time this was a common plant, as Nicholas Culpeper (1616–64) indicates in his famous *Herbal*. It was, he said, 'so well known to the inhabitants of every garden that I hold it needless to describe'.

Things have changed in the last 350 years. It is found in some towns, where it may be discovered growing at the base of old garden walls. Because it is an annual species, it often comes up in the same place year after year.

In the past the young leaves of the plant were put in salads, and sometimes added to wine cups.

---

is apt because of the silvery colour of the plant. Dust collects around it, which provides it with a substrate over which it can rapidly spread. The larger the area the greater the amount of dust collected, and so the faster the growth rate. Although most mosses reproduce by means of spores, the case of the silvery thread moss is the exception rather than the rule. Increase occurs when the tops of the leaves fall off, and rain water carries them away from the parent plant to new areas. Here they may start to grow if conditions are suitable. It seems possible that small animals moving about on the wall may also transport bits of the leaves to new areas.

Ferns may do well

Originally the only habitat for ferns would have been natural materials, say in the cracks between rocks, once enough soil had collected. Now these rock-loving ferns have taken advantage of the man-made environment. With a preference for a calcareous base most of the ferns have established themselves in the mortar in the cracks, because this provides the necessary mineral requirements. The distribution of species varies from area to area, but it is generally considered that as a country-wide species the maidenhair spleenwort (*Asplenium trichomanes*) is dominant. Perhaps it is not surprising to find that in some places it has the word 'common' added to its title. It reaches a maximum height of around 20cm (8in), with smaller specimens being no more than 4cm (1.75in) tall, and has leaves which are made up of between 15 and 140 oval-shaped *pinnae*. The spleenwort reproduces by means of spores, which appear between May and October.

Black spleenwort (*Asplenium adiantum-nigrum*), fairly common on walls, is likely to be found more often in the west of Britain. It gets its name from the long stalks which are black to purple in colour. Reaching a height of between 10 and 15cm (4 and 6in), its spores are ripe from June to October. It may also be found on hedgebanks, as well as in shaded positions on walls.

WORMWOOD (*Artemisia absinthium*)
A speciality in urban areas of the West Midlands, the wormwood is uncommon in many southern areas, including London.

It will grow to a height of 90cm (36in), and has distinctive silvery leaves. Its yellow flowers will appear in July and August. It has very bitter properties and because of this has been used to colour vermouths and to make absinthe. It was considered to have antiseptic properties, and earlier folk used to take it to 'rid the body of worms' — hence its name.

Larger ferns may colonise the wall
Given suitable conditions, larger fern species, like the harts-tongue (*Phyllitis scolopendrium*) and the male fern (*Dryopteris felix-mas*) will find a position on the wall. Although harts-tongue is more at home in woodlands, as long as there is plenty of moisture on the wall it will manage to establish itself. It has been noted particularly close to broken guttering, where there is an almost regular supply of water. Harts-tongue is unique as far as British ferns are concerned: whereas all other species have divided fronds, those of this species are solid. It is the heart-shaped base of the leaves which probably give rise to its name. A good specimen may reach a height of 60cm (24in) on a wall. Spores will be produced in July and August.

Other fern species, like polypody (*Polypodium vulgare*), rusty-back fern (*Ceterach officinarum*) and wall rue spleenwort (*Asplenium ruta-muraria*) may be found from time to time.

Flowering plants brighten up some walls
Flowering plants too may manage to find a space on the wall, plants such as wall rocket (*Diplotaxis tenuifolia*), wall-flower (*Cheiranthus* spp.), wall speedwell (*Veronica arvensis*), wall pepper (*Sedum acre*) and wall pennywort (*Umbilicus rupestris*). All these plants have acquired the tag 'wall' in relatively recent times, previously being found in natural cracks between rocks and boulders, like the other wall species.

The wallflower is a typical biennial garden species, which the plant breeder has developed over the years. However, as a naturally wild species, it is perennial by nature. Provided that there is enough material for food in between the rocks, the wallflower will gain a foothold, and flower year after year. Not native to the British Isles, it is thought to have come originally from the Mediterranean.

Pellitory-of-the-wall (*Parietaria judaica*) has an interesting alternative name of artillery plant. This is well deserved, when one considers its method of pollen dispersal. The male flowers possess very prominent yellow stamens. These parts are ensnared by the sepals. As the pollen ripens, the sepals bend and cause the stamens to arch. This means that they are under great tension. When something touches the stamen — perhaps an insect crawling over the wall, or even the wind — the filament springs into the upright position. As it does this it releases the pollen, which can be seen as a visible puff.

A member of the nettle family, and so related to the more unpopular stinging nettle, pellitory-of-the-wall is native to this country. Once established, the rooting system soon ensures that it will obtain the necessary nutrients as the roots seek out cracks and crevices between the main building blocks.

A species which ought to do well in the more exacting conditions which the wall offers is the wall pepper or stone-crop. The leaves are succulent, and this ensures the conservation of water. It gets the name 'pepper' from the taste of the leaves. When in flower the bright yellow star-shaped blooms make the species extremely conspicuous.

Some succulent leaves
Local people have always coined their own particular names for certain species of both plants and animals, and this is true of the wall pennywort. The 'penny' is a reference to the leaf-shape. In certain parts of the country it has come to be known as 'navelwort', a term used because of the dimple-like depression in the centre of each leaf. Like the wall pepper, it too has fleshy leaves.

The ivy-leaved toadflax (*Cymbalaria muralis*) comes from southern Europe, but was brought to our shores at the beginning of the seventeenth century. It quickly went wild and now it is found on walls in many parts of the country. Once established it spreads fairly rapidly, the trailing stems giving it the necessary means to spread. The flowers, a delicate shade of yellow or purple, will attract the attention of bees, which will visit them for the nectar which is found in the nectary. An interesting adaptation to wall-living is found in this species. After pollination the ripe seeds are held in capsules at the end of long, thin stalks. When ready they react to the light, so that they turn to face darker crevices. Once the seed is released it will hopefully end up in the crevices between the building blocks. If it does it will eventually start to germinate.

Small animals seek refuge

In some thicker, older walls, it may be possible to find mammals. Birds too will relish the shelter which many walls provide, and will probably find nesting sites in some of them.

For variety and numbers of species we have to turn to the backboneless animals – the invertebrates. Generally by far the most conspicuous of these animals are the spiders, or if not the animals themselves, then the numerous webs which they seem to be able to spin wherever there is a wall – be it around a garden or as part of a building.

To most people any small animal is an insect, and spiders are often mistaken in this way; but they have features which distinguish them from insects and other invertebrates. They have eight legs and the body is divided into two parts, while insects have three body parts and three pairs of legs. Another misunderstanding is that all spiders spin webs, but this is not true. For example, a wolf spider (*Pardosa amentata*), as its name suggests, hunts for food.

Spiders are especially noticeable from late summer through to early autumn. One of the most frequently encountered is the garden spider (*Araneus diadematus*). A web-builder, it constructs a very beautiful orb-web. Using spinnerets the female spider produces silk which she uses to make her 'home'. Since her food consists of unsuspecting flies, she will position her trap – the web – in a place which flies are likely to frequent. Incidentally, both male and female garden spiders produce silk when they are young.

Although more webs are likely to be found where there is some plant cover, even the bare wall will support at least one and often several different species. Older, weathered walls, which generally have numerous cracks and crannies, will be ideal places for spiders. Do not overlook new walls, however, because there is usually some place where a web can be constructed.

Undoubtedly the most beautiful structures are the

COMMON WASP
For much of the year the common wasp goes about its task unnoticed by humans. It is a useful friend of the gardener, taking a variety of larvae, which it carries back to its nest to feed to the grubs. It is in late summer, when the queen has left to seek out her hibernatory quarters, that the other wasps desert their duties, and seek their fill from autumn fruits.

orb webs which are made by other species. The two most common are *Zygiella* spp. and *Araneus umbricatus*. As long as somewhere can be found to attach the web, almost any wall will attract the interest of these species. There is, of course, another requirement: there must be some form of hiding place within easy reach of the web. Spiders are small, and the wall usually provides plenty of secretive nooks and crannies. The spiders will find holes in the building material or, in the case of older walls, beneath one of the capping stones or tiles.

Zygiella sp. is known as the telegraph spider, not because it builds its web on telegraph wires or poles, but because it is usually possible to see one long thread running from the corner of the web, leading to the whereabouts of the hidden invertebrate. Although it is an orb-web maker, the telegraph spider is not as particular as some other orb making species, and it is quite often possible to see one of these webs with part of it missing. With other spiders this generally means that the web is not occupied, but in the case of this species it is one of its trademarks.

The other common wall-dwelling orb-web maker is *Araneus umbricatus*. Related to the better known *Araneus diadematus* (the garden or cross spider) it is seldom likely to be encountered by day since it is strictly nocturnal, but a *gentle* poke at the web with a piece of grass may encourage it to emerge from its hiding place. The spider will most likely fold its legs around its body, and simply fall to the ground.

Like the telegraph spider, the sieve and comb spider (*Ciniflo* spp.) is also nocturnal, and webs are quite frequently encountered. As long as the spider can find attachment points when spinning its web, it will make do with almost any type of wall. The web is so dissimilar to the beautiful orb-webs. Almost carelessly constructed, it presents a generally untidy appearance. Sometimes the threads appear to have a bluish tinge; others are whiter.

Hidden species of spiders

Whilst it is possible to detect the presence of these spiders because of their webs, there are likely to be other species whose presence will only be detected if you actually come across them. Several species of spiders which catch their food by hunting may be found on the wall. As with the web-builders, the nooks and crannies in the wall are important to these hunting spiders. On a warm day you may be lucky to see them taking advantage of the stone-wall radiator, as they sun themselves. *Salticus scenicus*, the zebra spider, is a hunter rather than a web-builder. It has the ability to jump,

GARDEN SPIDER

The attractive orb-webs of the garden spider are woven almost everywhere throughout the town. Although they are to be found in every month, the garden spider becomes particularly conspicuous from about July or August. It is at this time that the spiders reach adulthood. Because of the unmistakable cross-marking on the back, this species is also called the cross or diadem spider.

and with eight eyes it can detect most movements, and so can take necessary evasive action. Its common name comes from the fact that it is black and white, the 'stripes' vaguely reminiscent of the zebra. With its shorter legs, the body of the spider is closer to the surface over which it is moving, and this makes it easier for it to creep up to unsuspecting animals. Moving up a wall and jumping could prove a hazardous activity for the zebra spider if it missed its foothold; but it has already thrown out a thread, which it has attached to the wall. Thus it can be seen suspended in mid-air as it endeavours to regain its foothold.

SPITTING SPIDER (*Scytodes thoracica*)
A strange species which is confined to southern parts of England, the spitting spider traps its prey literally by spitting. Usually living in warm out-houses, including sheds and greenhouses, the spider makes its way very slowly over the walls in search of its prey. When less than 10mm (0.3in) from its target, it lifts up the front part of its body, and expels two sticky threads from its jaws. Should they hit the prey—and the spider is not thought to miss very often—it will be held by the threads. A closer examination of the captured prey will show that the threads are not straight, but arranged over the prey in a zig-zag way. The spider is able to do this because it moves its jaws very quickly at the same time as it is spitting.

Another hunter

The other frequently seen hunting spiders are the wolf spiders. Using similar tactics to those employed by the much larger carnivorous mammal from which it takes its name, the wolf spider makes furtive advances on its prey. Not needing webs to catch its food, it nevertheless has silk producing spinnerets.

Conditions for bees and wasps

Masonry bees and wasps may make use of the wall if conditions are suitable. These, unlike the bees and wasps with which we are probably most familiar, live solitary lives: they only produce nests for small numbers of offspring.

A mason wasp (*Odynerus parietum*) is not quite so distinctively wasp-shaped as the common wasp (*Vespula vulgaris*). It lacks the noticeable waisted area. But the mason wasp can be identified by its colouring. It has black head, thorax and body, with the typical yellow banding breaking up the general body colour. The thickest of the bands is nearest to the thorax; the thinner bands towards the rear.

ANGLE SHADES

With its wings folded at rest, the angle shades could easily be mistaken for a leaf. It is not unusual for the insect to pretend that it is dead if it is disturbed during its daytime resting periods. Both these features are useful for protection.

SILVER-Y

It is from its silver metallic markings that the silver-Y moth gets its common name. Although it occurs in most areas in the British Isles from time to time, it is a migratory species, and it may be common in one year and seldom seen in the following.

Having found a suitable position in the wall, the wasp laboriously sets about improving the entrance and building a nest. Using soil mixed with its own saliva it makes its own mortar. Within the nest the female will produce her cells. These she usually arranges end to end, and in them she will place small caterpillars which she has caught. Although she makes sure they are immobile and unable to escape, they do not die, and as each wasp egg hatches into a larva it will have a supply of fresh food!

There are special mason bees (e.g. *Osmia* spp.) which also operate in a similar way to the mason wasps. There are several species, but one which is seen quite frequently is the red osmia mason bee (*Osmia rufa*). It has acquired the tag 'red' because of the distinctly reddish hairs which cover the abdomen.

These bees, like their relatives the honey bees, are especially active on sunny days. They seem unable to make a flight path directly to the wall, but instead flit around for a while, perhaps even going to one or two holes, before they are satisfied that they are entering their own domain.

In need of protection

One of the most common snails found in the British Isles is the banded snail (*Cepaea nemoralis*). It has a variety of names, including brown-lipped, hedgerow and grove snail. Naturalists have suggested that the 'typical' banded snail has five bands, either brown or black in colour, but in practice no fewer than eighty-nine different combinations of bands have been discovered in various parts of the country. In some cases the bands are fused together, which, to the uninitiated, gives the impression of a different species.

The wall provides the banded snail with a much-needed daytime resting place. Like other snails, it is a nocturnal feeder, even though a shower of rain will entice it to leave the security of its diurnal sleeping quarters. It will return consistently to the same spot after it has been out to feed. Although it treats stinging nettles as its favourite delicacy, where these are not found it will make do with substitutes. The *radula* – in effect a tongue – consists of a large number of rasp-like projections, and these are responsible for tearing away the leaf tissue. As they wear out, they are replaced.

The banded snail in its turn provides a plentiful supply of food for many different animals. It is often the shells of this species which will be found around the thrush's anvil, but field mice, rats, hedgehogs, voles and even rabbits feed on them too. Although they are part of the song thrush's favourite food it does not seem to be the bird's main source of nourishment, and snails will only be eaten at certain seasons, particularly in winter and spring.

Like earthworms snails have both male and female sex organs, but mating is necessary, with the result that each animal will be fertilised. Mating is generally considered to take place in the morning. The slime trail which the snail leaves is sought out and followed by another snail. At the encounter they rear up, each biting the other. A love dart is then ejected, first by one snail and then by the other. Mating follows, and it is not unusual for this to go on for several hours.

Egg-laying will follow. The first will usually be deposited in May, and other snails will not lay theirs until August. No eggs will be laid if conditions are dry, as they would dry out. It seems that in a good season as many as four clutches may be laid. The number in each clutch will vary according to the type of banded snail. Those with five distinct bands appear to lay the most.

As egg-laying approaches the snail prepares its nest. To start with it pushes its muscular foot into the ground, and then by moving its body manages to produce a round hole. Here the eggs are laid. It takes up to three days for the process to be completed, with an interval of between 10 and 30 minutes between each egg being deposited. The last egg laid, the snail will cover the hole.

Once they have hatched, the young snails, extremely small replicas of their parents, have to dig their way

out of their 'prison'. Having released themselves they now make off in search of their first food. The next two or three years – which is the time taken to reach maturity – will be a very trying time for these snails. Before this stage is reached, many, perhaps the majority, from each clutch will have died.

During dry conditions in summer, the snails seal the outside of the shell, and spend the period resting, returning to normal activity when things improve. The cold weather in winter would probably kill off most snails if they did not hibernate. Again, as in summer, they seal the mouth of the shell with an epiphragm. This may consist of a number of layers, and will do much to prevent the snail's death in the cold winter weather.

Small birds seek holes in walls

Although no exact numbers are available, the wren (*Troglodytes troglodytes*) is quite a common bird in towns. It comes to gardens, not only to seek food from the bird table but because such places offer it some shelter and seclusion. Because of its small size the bird often manages to makes its nest in a hole in a wall. The male builds several nests, and once he has persuaded a mate to take a fancy to one of them, she will put her own stamp on it by lining it with feathers. It is quite common for the wren to produce two broods. The cock has been known to take over the first brood, housing them in one of the nests which he made earlier. Here he will care for them whilst the female is engaged with dealing with the second clutch of eggs. By the time that these hatch, his first family will be off his hands, so to speak, and he will be able to concentrate on feeding the second brood.

Most of our common birds have a variety of superstitions or folklore associated with them, but none more so than the wren. Perhaps one of the strangest is the fact that the wren was considered the 'king' of birds. That so small a bird should end up with so grandiose a title is remarkable; but there are many instances of this royal epithet. In France the bird was known as 'king of the birds' (*roi des oiseaux*), and the Latin was 'little king' (*regulus*). There are many similar titles in other languages, so the small bird's fame was quite widespread. According to a French legend, the bird was actually present in the stable in Bethlehem when Christ was born, and so it is known as *poulette de Dieu* (God's chicken).

The house martin (*Delichon urbica*) needs the eaves of houses and other outbuildings as a place to attach its well-known nest. Because of this, it is considered a town bird, and it does occur even in the heart of some of our largest cities.

As with some other species, house martins have made use of man. Initially they were known to build their nests in caves and in cliffs, but as man has provided them with alternative sites, they have taken advantage of these.

The familiar mud nest, with roots and grass to help bind it together, is fixed under the eaves of a house. Once completed, the interior of the nest has a substantial lining of feathers.

When the eggs are laid the usual clutch consists of between three and five eggs, which appear between May and August. Both parents will spend the next 14 or so days incubating the eggs, and both will feed the nestlings once they are hatched. Generally there will be two broods, and a third has been noted in some seasons.

The nest is not necessarily abandoned once the young have hatched and flown. The birds from the current season's brood will often use the home as a nightly resting place. It seems that one nest will attract not only its earlier occupants but other birds as well.

Life under the hedge

Hedges in towns support a wide range of both plants and animals. Although we tend to see various wildlife habitats as catering for only one species, this is far from the case. For example, the hedge may support thrushes and we may discover banded snails living in the base. The snails will probably provide a valuable supply of food for the thrushes.

There are many different kinds of hedges. In very old properties, particularly on the edge of towns, there may be ancient ones. These could have been left when the property was originally erected, and so be even older than the house. Alternatively the plants may have been put in when the house was built.

Many hedges are of course much newer. Where the vogue is to have a natural barrier as quickly as possible, quick-growing conifer species are often planted. Although these will provide some refuge for wildlife, their value is a good deal less than where mixed hedges, consisting of deciduous species, are found.

Cover for mammals

If there is plenty of cover it is likely that several species of mammal will find refuge in the hedge. Because it is one of the most common and widespread species of mouse, we should perhaps expect to find here the long-tailed field mouse, also known as the wood mouse (*Apodemus sylvatica*). Even in cities it is likely to be found in the hedge, and it may turn up in the garden as well. The total length of the animal is about 17cm (7in): of this about 8cm (3in) is tail. If conditions are suitable and there is plenty of food, the wood mouse will breed prolifically. There will be between five and nine youngsters in a litter, with seven being normal.

The female wood mouse digs a hole for the nest which she makes with grasses, moss and dried leaves.

Adaptable as far as food is concerned, the mouse feeds on a wide variety of material, including snails, birds' eggs, nuts, bulbs, vegetable matter from many sources, seeds and fruit. It is not a hibernating animal, but instinctively starts to form a winter hoard in autumn. If there are disused nests of blackbirds and thrushes in the hedge, the field mouse will use these to collect together a store of food. It is a fine sight as it sits in the nest tucking into its food.

If there are berries and other food in the hedge, it will find most of its needs here. However, ever adventurous, it is not averse to taking foodstuffs from the garden, and it is for this reason that it is not always welcomed as a visitor.

The urchin visits too

Although most people are familiar with the hedgehog (*Erinaceus europaeus*), in fact these creatures are seldom encountered because of their nocturnal habits. They are frequently found in towns, where garden and other hedges provide them with a hiding place. It is also known as the hedge-pig, a reference to its snorting and grunting as it goes about its nocturnal food-hunting expeditions. The word 'hedge' features in both names, and this is a reference to its diurnal resting place.

A friend of the gardener, until the advent of modern insecticides, the hedgehog was a natural pest controller, feeding upon a veritable treasure house of grubs, beetles, and larvae in the garden. Although it will still find an abundant supply in most areas, there are places where gardeners have used so many sprays that the hedgehog's food source has been eliminated.

As dusk creeps over the townscape, the urchin will stir from its resting place, to embark on the all-important task of food collecting. With a diminished light source, its highly developed senses of smell and hearing are particularly important. Shuffling and snorting its way around the herbaceous border, it is likely to find all that it needs. Although it is classed as an insectivore, the hedgehog does take other food.

In late summer and early autumn, when fruit has fallen to the ground, it will seize this as an acceptable alternative to its normal 'flesh' diet. Both young and eggs of birds which nest on the ground – especially in the countryside – will be vulnerable to our friend. Although not a regular feature of its diet, it will take other animals, including frogs, mice and rats. The occasional snake may be food for some hedgehogs.

Where the base of a hedge is thick and provides good cover, the hedgehog will have its breeding nest, made from leaves and grass. It may produce young between May and September. In a good season, and where the first mating is early, a second litter may be born. Between three and seven piglets are born at a time. At birth the young hedgehogs are helpless: they are pink and blind, and the body, later covered with strong protective spines, is now particularly vulnerable, having only a sprinkling of pale soft spines. A couple of weeks after birth, these spines begin to harden. The female will provide milk for the young for about a month, until they are ready to search for their own food. In the first part of the food-hunting initiation ceremony, the parents will begin by taking the youngsters on their nightly excursions. A family of hedgehogs being led by an adult is not an unusual sight. A sign that a hedgehog has been on your patch are the tell-tale cigar-shaped droppings.

The winter period is spent at least in part in a state of hibernation. Older animals tend to retire early – perhaps in October. Younger ones postpone their eventual capitulation to sleep until later, perhaps because they need longer to build up a supply of stored food than do older animals.

The winter hibernatory home has to protect the hedgehogs for several months, and more care is taken with its preparation than with the summer sleeping

HOUSE MARTIN
House eaves will attract the attention of the house martin as it seeks a site for its nest. It will repeatedly return to the same eaves even if the nest is destroyed by the householder. As towns have become cleaner the house martin has moved into inner city areas. The bird returned to London in 1966 after a lapse of seventy-seven years.

quarters. The nest will be larger than the nursery nest. Weatherproofing is important, and the hedgehog uses a combination of moss and leaves in an effort to keep out the elements. This also helps to ensure that the nest is insulated, so that the interior temperature remains fairly constant.

Although hibernation implies sleep throughout the winter, the hedgehog is likely to wake up and leave the nest from time to time in search of food. This will usually happen when the temperature rises.

In some areas it is not unusual for several hedgehogs to live together in one area. As with some other mammal species there seems to be a status system. In gardens where several hedgehogs have come to the same bowl of milk to feed there seems to be a senior-to-junior drink order. Whether the hedgehogs are all from the same family, or whether they are unrelated, has never been established. Feeding hedgehogs regularly tends to 'tame' them, and instead of running away they will stay.

In spite of the temptation, it is not always the wisest thing to pick them up. The spines – in effect modified hairs – may cause the uninitiated some problems, but these could be minor when compared to the mammal's abundant visitors. The whole body is covered with fleas. These do not appear to do the hedgehog any harm, and they will die quickly if they leave the mammal for a human being!

A place for birds
Hedges are important for birds, and where town hedges are suitable it is likely that several bird species will manage to use them. For some it might be a hiding place, for others a source of food, and for yet more a place to nest and rear young.

Blackbirds (*Turdus merula*) are familiar to most people, whether they dwell in town or country. It is probably fair to say that the male, with his bright yellow bill and jet black plumage, attracts more attention than the female, which, with dark brown plumage and lacking the male's bright bill, is much less conspicuous than her partner.

Perhaps the keynote of the blackbird's lifestyle is adaptability. Not only can it nest almost anywhere – and so besides being found in hedges it will occur in other urban habitats – it is able to survive on a wide-ranging diet. This is probably why, along with the chaffinch, it is probably Britain's most common bird. Whereas some birds have suffered as man has expanded his operations, the blackbird seems to have managed to live successfully alongside these new developments.

The increase in the blackbird population is probably due to the fact that it breeds a number of times during each season. Two broods is certainly the norm, and

three is not considered uncommon. In a particularly good season mating can take place more often.

Last year's youngsters seem to be eager to get on with mating and nest-building. In fact their enthusiasm is such that nests will probably be built before the leaves have appeared on the hedgerow shrubs, and their nests are particularly conspicuous at this time of the year. It is the female which undertakes the nest-building operation. Starting with dried leaves and grasses she will mix these with mud, creating a bowl-shaped structure. The first eggs will be in the nest in March and in some seasons they may be found until July. Some 13 days after she first sat on the eggs, the young will hatch. Although the cock bird does nothing to help with incubation, his attitude changes once the young are in the nest, and he brings a constant supply of food. A member of the thrush family, the blackbird's two common breeding relatives are the song and mistle thrushes.

For the urban dweller who is awake to hear the dawn chorus – and it is worth making the effort now and again – it is probably the blackbird's melodious voice which comes through most often.

At home in the hedge
The blackbird's relative, the song thrush (*Turdus philomelos*), will also be equally at home in the town. Here it can find nesting sites in hedges as well as in bushes and trees. Although the blackbird has managed to gain an increasing foothold in towns, the reverse has been true as far as the song thrush is concerned. Numbers in the built-up areas appear to be on the decline. The reason is not certain, but it is possible that since it feeds on much the same material as the blackbird it is losing out to the increasing numbers of its relatives. The blackbird has a bolder approach to feeding than the song thrush, and this may be another reason.

Ready for nest-building, both mates will set to with a will to produce a large structure. Dry grass and dead leaves are employed and the interior is given a coating of mud. Eggs, usually three to six in number, are laid from March to July. Incubation is carried out by the female, lasting for about fourteen days. Both set about feeding the nestlings, searching out a wide selection of insect larvae, as well as earthworms and snails removed from their shells.

The word 'song' in the bird's name is appropriate. Starting as early as January, the melodic song fills the quieter parts of towns with beautiful sounds. Once in song the thrush will continue its choruses until July. It then usually keeps quiet for a couple of months, but once late summer is here it starts up again in September, and will go on until November.

DUNNOCK OR HEDGE SPARROW

Originally a bird of wooded areas, the dunnock (or hedge sparrow) has adapted to built-up areas as more and more woodland has disappeared. On the periphery of towns the dunnock may find that the cuckoo has chosen its nest for egg-laying. The dunnock then has to feed the young cuckoo, a feat which takes considerable effort.

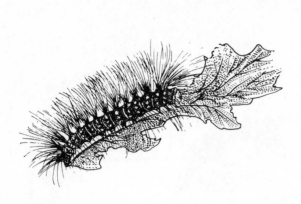

GARDEN TIGER

The garden tiger is one of the most noticeable of the British moths, both as a caterpillar (above) and in the adult form (right). Brilliantly coloured with black and red, the adult exhibits the clearest possible warning for other animals to stay away. Such bright colours—warning colouration—indicate to would-be predators that the creature does not taste good. The caterpillar is commonly known as a 'woolly bear', because the body of the larvae is covered with a thick coat of hairs. The caterpillars feed, but they do not become fully grown in the first spring and sleep through the winter. Once they emerge from hibernation they feed again until fully grown and ready to pupate.

Having a preference for snail flesh, the song thrush is well-known for its anvil. This is a suitable stone, which is used in the extrication of the soft-bodied creature from inside the harder, inedible shell. Striking its victim against the anvil, the thrush leaves behind a collection of broken shells, a sure sign of its continuing activities. Earthworms, as well as insects and other larvae, will also be taken. This sort of food is not so readily available in the winter, and so the thrush changes its diet, seeking out seeds, berries and fruits.

Man's favourite bird

It is the robin (*Erithacus rubecula*) which is undoubtedly man's favourite bird, particularly in winter, when it comes closer to human dwellings. Originally living in woodlands, robins have taken advantage of man's situation, as he has changed the face of the countryside. With cock and hen birds similarly marked it is difficult for the layman to distinguish between the two.

The endearing behaviour of the robin in winter certainly belies its aggression when defending its territory. This area is carefully staked out and patrolled and woe betide any intruders! To warn would-be trespassers, the bird keeps up a repetitive call, the

purpose of which is to determine its territorial boundaries. During the winter months the hen bird also lays claim to her own territory; but both will probably desert their areas as a temporary measure should the need for food become overwhelming.

Having found herself a suitable nesting site, the hen bird will produce a deep cup-shaped nest, using a mixture of dead leaves, grass and moss. She may place a roof over it when it is in a cavity. As far as some robins are concerned, the breeding season may start as early as March; for others it may go on until June. A single clutch consists of between three and six eggs. With a base colour of white, there is a variety of spots and blotches of a red-brown colour, breaking up the surface. The cock leaves his mate to incubate the eggs. but does take his turn with feeding the nestlings. Two weeks after they hatch they will be ready to leave, and provided it is early enough in the season this leaves the pair free to raise another brood.

Not strictly a sparrow

The hedge sparrow (*Prunella modularis*) or dunnock, although labelled by our ancestors as such, is not strictly a sparrow. Any small bird was once given the title 'sparrow'. However, the name 'hedge' is well deserved, as the bird spends much of its life moving about among the hedgerow vegetation, where it is ever-watchful for a supply of food. Its bill gives away its eating habits. Longer and more slender than that of the familiar house sparrow, it is shaped for dealing with soft material. A wide variety of insects and their larvae, ever-present in the hedgerow vegetation will provide it with food. If these are difficult to come by, the bird may turn its attention to spiders and earthworms. As with many other insectivorous species its diet changes in winter. Insects are difficult, if not impossible, to come by at this time of the year, but seeds are usually available as an alternative food source.

The bird has acquired its other name, dunnock, from the colour of its plumage – dun coloured. In the southern counties of Britain the bird is the target for the 'lazy' cuckoo. The female cuckoo will lay her own egg in the nest, leaving the sparrow to incubate it. After hatching the nestling will push out the other bird's eggs. This may occasionally happen on the edge of towns. The dunnock as well as the cuckoo may be absent from the more densely populated and heavily built up areas, especially where hedgerows are less common. Nevertheless it does often go about its

PRIVET HAWK MOTH
The privet hawk moth emerges in June, and after the adults have mated the female lays her eggs on the leaves of privet. Although it is a nocturnal species, it is often encountered during the day when it is disturbed from resting on fences and walls. The young caterpillars may be encountered feeding on privet, although they are often passed over. If the creature is alarmed it assumes a defensive stance, in which the front end of the body is raised.

business unnoticed, its drab feathers being well camouflaged against the vegetation.

When seeking out a site for its intricate nest, the hedge sparrow will be looking for a hedge which has good cover just above the ground. Where hedgebanks provide an adequate refuge, the dunnock may start to build its nest from a combination of rootlets and fine twigs. Once the basic structure is completed the interior will be lined with hair or wool, ready to receive the bright blue eggs.

The female hatches the eggs for two weeks, and with the birth of the chicks the cock decides to help his mate. Together they bring what seems to be a continuous supply of food to satiate the appetites of up to five nestlings. Within twelve days or so they are ready to leave the parents, and they can, if the time is ripe, concentrate on the rearing of a second brood.

Large numbers
By far the greatest numbers – in terms of both population and variety of species – of animals to be found in hedges are the invertebrates. These are the animals without backbones, and include many of the familiar insects, but also slugs, snails, and spiders, as well as the thousand and one other species which may

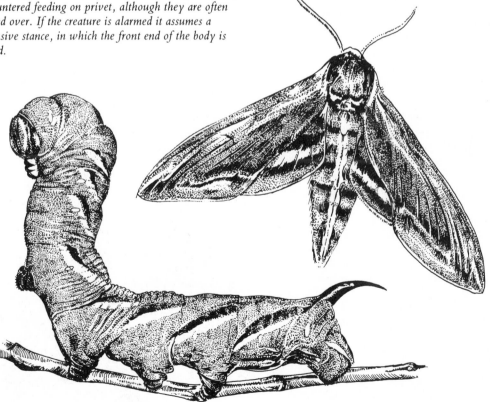

be found. As far as animals are concerned – large and small alike – the hedge is undoubtedly most active from spring through to autumn. For most species the earlier part of this period marks a time to find mates and reproduce their own kind.

The succulent leaves on the shrubs, trees and herbs of the hedgerow provide a rich supply of food for a variety of species. Of all the invertebrates present – and it has been estimated that Britain has 20,000 species of insects alone – none is particularly big. Many will be encountered from time to time in the built-up areas, although most go about unnoticed.

Butterflies and moths

Perhaps the most conspicuous of the invertebrate animals are the butterflies and their many night-flying relatives, the moths. The largest night-flying insect native to Britain is the privet hawk moth (*Sphinx ligustri*). With a wing-span of 10cm (4in), the nocturnal species is often associated with towns, and as its name implies it is frequently encountered on privet (*Ligustrum* spp.). The moths appear on the wing in late May. In June the female lays her eggs on the leaves of privet, but if she is unable to find this species she will turn her attention to lilac (*Syringa spp.*). Although one would expect to find the caterpillar – it may be up to 10cm (4in) in length – it quite often goes about unnoticed by the people around it. If the caterpillar is disturbed it will raise the front end of the body in a threatening manner. Once it has reached full size, it will move down to the soil, where it will bury itself. Below the surface it will change into a brown pupa, remaining in the soil until the following spring.

Moth plagues

Where hawthorn has been used for hedging, the small ermine moth (*Yponomeuta padella*) may be found. Indeed in some seasons the caterpillars reach plague proportions, when they will strip the leaves off the hawthorn leaving a bare hedge. When at rest the folded forewings exhibit a bright white colour, which is broken by a series of irregularly spaced black dots. Once the caterpillars have hatched they will form a communal web. They will need to go out to feed regularly, but the home serves as a useful shelter.

Hawthorn also provides a supply of food for the larvae of the vapourer moth (*Orgyia antiqua*). It does have alternative supplies, and if it is not found on hawthorn it may feed on oak, rose, lime or birch. The caterpillar should not be handled, since its numerous protective hairs exude a liquid which causes some people to come out in a rash. Although the males are capable of flight, the female lacks this ability, being equipped with extremely small wings. She remains on the cocoon after she emerges, and the male is able to

LIME HAWK MOTH (above)
The numbers of lime hawk moths vary from one season to the next and there is also a great variety of colouration and markings. The female lays her eggs on elm as well as lime. The caterpillar has seven distinctive oblique stripes, which can be seen on either side of the body.

POPLAR HAWK MOTH (below)
The poplar hawk moth is a frequent visitor to gardens, where it comes in search of nectar. Ready to lay her eggs, the female searches out either poplar or willow trees. Both these species are grown in gardens, and the female may find a suitable place there to deposit her green oval eggs.

find her because of the scent which she gives out.

Another hawthorn eater is the small eggar (*Eriogaster lanestris*). The female lays green eggs in batches, and she then covers them with hairs from her body. Once the eggs hatch the caterpillars set about making a communal web. A strange aspect of the life of some small eggars is that they may stay in the pupal state for up to three years before they turn into adults.

The yellow tail moth caterpillar (*Euproctis similis*), strikingly marked with red and black, is another species which may cause irritation of the skin. A common species, both moth and caterpillar may occur in the hedge, since they have a number of different food plants. It is the female moth which has given rise to the species' name. At the tip of her abdomen she has a conspicuously large tuft of hairs. These are important: after laying her eggs she uses these to cover them.

The caterpillars which emerge from summer eggs do not finish feeding in the first year, but will hibernate for the winter. Emerging in the following year, they will continue to feed. Once they have changed into adults, they will mate, and the females will lay their eggs.

Beautiful butterflies

Larger than the moths and more often encountered because of their diurnal activities are the butterflies. Where there is a plentiful supply of food many different species may be found. Not necessarily confined to the hedge – or to the garden – nevertheless they are very welcome visitors to the urban scene. Unlike moths, few butterfly species actually lay their eggs on hedgerow shrubs. One of the earliest butterflies on the wing is the sulphur-yellow brimstone (*Gonepteryx rhamni*). Having spent the winter hibernating in the open, its fragile body perhaps pressed against the trunk of a suitable tree, it is not surprising that it should be stirred by the first warm rays of the sun, and awake now from its winter slumbers.

In spite of its early flight, it will have to wait until its food plant – buckthorn – is in leaf before it can lay its eggs. Mating over, the female will deposit her eggs in either May or June on the underside of leaves. The caterpillars will feed for about two months, and then pupate, so that the adults will emerge towards the end of July or early in August.

Food for caterpillars

The hedge brown (*Pyronia tithonus*) or gatekeeper, as both names imply, is to be found along and around hedges in July. Providing that suitable grasses are available the female will deposit her eggs. Grasses also form the food source for other species of browns. Three species of butterflies rely on stinging nettles for food – the peacock (*Inachis io*), the small tortoiseshell

(*Aglais urticae*) and the red admiral (*Vanessa atalanta*). Any of these may find suitable plants around hedges.

Hedgebanks for wasps

Of course the variety of insect species which will be found in the hedge varies from place to place. Where a suitable hedge is available the common wasp (*Vespula vulgaris*) may build its nest, often to the abhorrence of the inhabitants of any nearby buildings. Where hedges occur on hedgebanks this may be a suitable site for the wasp's nest. The queens, having mated in the previous summer, hibernate, thus avoiding the cold and damp. With the arrival of spring, the queen common wasp will emerge from her winter's sleep, and will seek out a suitable place for her nest. She will need to lay her eggs as soon as she can, to ensure that the first members of her brood will be born. Keep watch from about the beginning of April to the end of May for wasps. You will notice that they are rather larger than those which you will see later in the garden feeding on fallen fruit. These are the queens. Once they have found a suitable hole in the ground nest-building begins.

Just as a bird needs nesting material so does the queen wasp, but unlike them she needs wood. Using her strong mandibles she will tear off wood shavings from posts and fences, which she will then have to mix with saliva to form wasp paper, which is brown in colour and rather brittle. Starting at the top so that the nest is suspended, she will begin her chores. Working consistently and conscientiously she soon has a number of cells arranged next to each other. These are ready to receive the first eggs, and because the opening is at the bottom it is necessary for her to cement the eggs in position. Developing quite quickly, the eggs soon hatch and now the queen has yet another job as she tends her offspring – the larvae. She has to collect a constant supply of caterpillars and other insect larvae, which she first chews up before offering them to the larvae. Regular feeding ensures that they grow rapidly.

Ready to pupate, they seal up the end of the cell. This silken pad is made from threads which the larvae produce in glands under the head. This is an important stage in the life cycle of the wasp, for the larvae will quickly pupate and make the change to adult. Usually no more than five weeks after she laid the first eggs, the queen will have some new helpers in the nest. These are the workers. Although females, they are not capable of egg-laying, and as their name implies, they have work to do. Two tasks await them. First, the nest needs to be increased in size, so that the queen can continue to lay her eggs. Second, as the increasing number of larvae hatch they will need a constant supply of food. This the workers provide, leaving the nest to search out insect larvae. Relieved of her irksome chores, the queen now devotes all her time to the task of egg-laying.

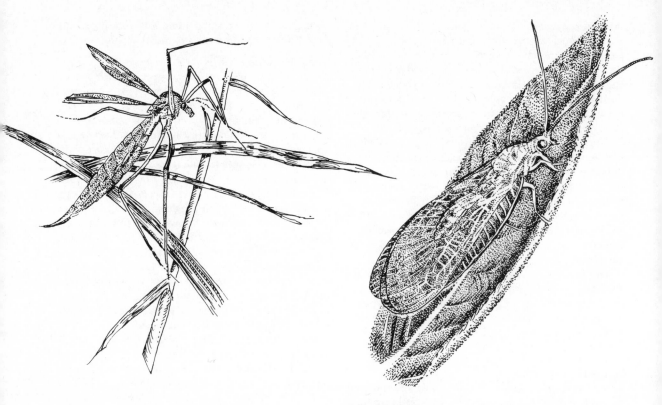

CRANEFLY

The cranefly occurs almost universally where there is grass. The adult lays as many as 300 eggs in the ground. The young stage, called a leatherjacket because of its appearance, feeds on the grass roots. When nights are warm and humid it comes to the surface and nibbles the base of grass shoots.

The rate at which the size of the colony increases is spectacular. By the end of the season a single nest may contain as many as 5,000 individuals. There is yet one more important event. Arrangements have to be made for next year's wasps. Soon drones and young queens emerge from the eggs. Once they have made the change to adults they will leave the nest. The young queen needs to mate before adverse weather comes, and so this process is performed once they leave the nest. Drones from the same colony or from another may be selected for mating, so that the queens are pregnant before they go into hibernation.

Once the drones have managed to fertilise the queen, their job is complete, and so is that of the other occupants of the nest. It is only the fertile queens which, because they hibernate, are able to withstand the variations in temperature which will occur during the winter months.

The care which the workers have lavished on the large number of wasps in the nest changes as soon as the

LACEWING

The delicate lace-like wings from which the insect gets its name carry the lacewing from one plant to another when it is disturbed during the hours of daylight. Seemingly fragile, the insect has an excellent appetite, aiding the gardener as it feeds on harmful pests such as aphids.

last queen has made her exit. Their task is finished, and they make sure that any remaining larvae, no longer necessary, are killed. Then they desert the nest to meet their deaths as the colder autumn weather descends upon the countryside.

In the countryside the most common species used for hedges is the hawthorn. In built-up areas, although it is used, other more decorative species generally take precedence, and yet with its superb snow white flowers in spring, and luscious red berries in autumn, hawthorn is also useful in the garden. The flowers, generally called may by country folk, will usually start to appear around the middle of the month after which they are named. Because they appear in bunches they give the hedge a beautiful patchwork effect, with the white flowers intermingling with the green leaves. The bright flowers attract various insects, and pollination follows. The flowers will eventually disappear to be replaced in autumn by the haws which have developed over the months. Brightening up the autumn scene,

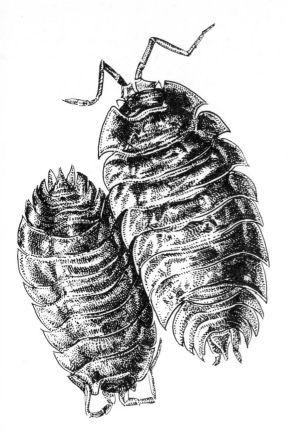

WOODLICE

Woodlice are nocturnal creatures of damp dark places. If they stayed in the open for very long they would lose water from their bodies and die. They find it safer to come out at night in search of a variety of decaying vegetable matter. They are extremely widespread and are found almost anywhere where it is damp.

they will also provide a good supply of food for hungry autumn birds. Mammals, too, such as the field mouse, will relish their nutritious and succulent flesh.

If you have hawthorn in your garden you may make the acquaintance of the hawthorn sawfly (*Trichiosoma tibiale*), yet another species which feeds on this common hedging shrub. Although both larvae and adults will probably occur in many hawthorn hedges for much of the year, they are difficult to discover. The larvae are well camouflaged, having a green skin which affords them protection on their food plant. Adults are seldom seen; but when the deciduous hawthorn has lost its leaves in autumn, the cocoons which house the pupae are revealed. The adult will feed on the hawthorn flowers when they are out, but seldom make prolonged appearances.

Wild flowers

Unfortunately gardeners are too tidy, and they regularly clean out the bottoms of their hedges. Yet those who come to favour a wilder hedge soon come to realise that a whole host of plants may establish themselves. Those which arrive owe their presence to some form of dispersal. They may have arrived unintentionally – at least as far as the carrier is concerned – in a bird's dropping; or perhaps they were transported by a visiting animal. Others may have been deposited courtesy of the wind.

Even at the base of most hedges, however, tidy the gardener is, there will usually be some wild flowers trying to find a niche. Where the hedge divides one garden from another the flora will undoubtedly be limited. Even in areas where modern buildings have sprung up, old hedges are often retained as boundaries. It is these, where they once surrounded a field, which often have an interesting herb composition.

One of the strangest plants which often grows in hedgerows is the wild arum. It has a variety of other names, including lords and ladies, jack in the pulpit and cuckoo pint, as well as a few others which you might like to discover for yourself!

The initial impression is that the cuckoo pint does not look like a typical wild plant. Above the green stem there is an area which distinctly bulbous. Inside this are both male and female flowers. Other male flowers appear above these in a part of the plant which takes on a purple tinge. Club-like in outline, this is called the spadix. Surrounding this is a cloak-like part called the spathe.

As the flower matures it does not endear itself to passers-by, because it gives off an aroma which has sometimes been described as resembling strong urine. If man is not attracted by its scent, however, there are flies which do not find it so intolerable. Nature is again planning a strategy which ensures continuation of a species.

Certain species of flies find the foul smell irresistible, and they crawl down the spathe. As they progress they pass by hair-like structures. Continuing on their journey they make their way past both true male and female flowers. Inside they do the job for which they have been enticed. Crawling about in the base of the cuckoo pint they inevitably make their way over the female flowers. They have made sure that some pollen has been deposited on the female flowers. Although they may endeavour to leave, their escape route is, for the present time, blocked. Above the true male and female flowers, the downward-pointing hairs prevent the flies from making their escape. Imprisonment is not permanent. Once the flies have crawled about for long enough they ensure that all the female flowers have been pollinated. This over successfully, the hairs wither and the flies can make their escape. This is a remarkable method provided by nature to ensure pollination.

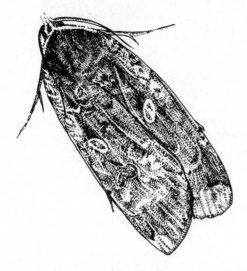

YELLOW UNDERWING (above)
Many people will be familiar with this species, even if they do not know its name. The yellow underwing is very common, and is one of the moths which comes into houses, especially during late June and early July. Although basically nocturnal in its habits, it may be seen during the daytime when its resting places are disturbed. Because the caterpillar eats many different kinds of both wild and cultivated species, it is common in many gardens.

MAGPIE MOTH (below)
The magpie moth makes its debut in July and is a common visitor to gardens. The adult has white, black and yellow-orange markings on its body, but there is an extremely wide variety of patterning.

So strong is the smell that the flies are attracted to other cuckoo pint flowers, and they seem undeterred by their short-term imprisonment. Whilst in the base of the cuckoo pint they are also provided with food, which enables them to survive. Once they have escaped, they make their way to other wild arum plants which might be around, and the whole process will be repeated.

Pollination accomplished, the berries begin to develop. At first they are green, and as the spathe withers they are revealed. Ultimately they take on a bright orange colour, and look particularly attractive. To humans who might be lured by their bright colour it is a warning: stay away, the fruits are poisonous. The poison, however, does not affect wild creatures which come for food, and in the process helps to disperse the seeds inside the berries.

The hedge provides a means of support for various climbing species. Ever eager to grow bigger to obtain more sunlight, to make more food, species like bindweed and travellers joy may seemingly smother the other hedgerow plants. It is said that travellers joy was given this name by the famous sixteenth-century naturalist and herbalist called Gerard. He was so taken with the plant on his travels, especially in the autumn when the plant is particularly conspicuous, that he indicated it gave great joy to travellers. However, Gerard probably travelled relatively little in his day when compared with modern voyagers. Although found growing on various soils, the plant is at its best on calcareous soils.

When the flowers appear they are greenish-white and do not produce nectar, which would normally attract insects. Nevertheless bees and flies do pay the flowers attention, since they find the abundant supply of pollen acceptable. Once they have been pollinated the tip of each stigma (the female part) begins to develop. It is hairy, and as it develops it exhibits this hairy appearance as the seeds mature.

It will be April or May before the seeds start to germinate. Once they start to grow and the first leaves have become established, the plant will start on its vigorous upward climb. Using the stems of other already established plants, it twines its stem clockwise around these. The leaves become tendrils, which enable them to cling on. As growth continues apace the once fragile stem takes on a woody nature, and the bark splits and falls off in long strands.

Both black and white bryony – which belong to different families – may find a suitable spot in the hedge. The flowers, like those of the travellers joy, are small and greenish-white. They can be seen on the plant in May and June. Fertilised, the berries start to develop. Soft and scarlet when ripe, they become particularly noticeable once the leaves have fallen from the climber. Animals will find them attractive, but humans should avoid them since they are especially poisonous.

The white bryony, which has a relative in the cultivated cucumber, has a longer flowering period. The first blooms appear in May; the last ones not until September. The hedgerow should provide the white bryony with adequate support. Its tendrils coil around a variety of other hedgerow plants strong enough to support it. Virtually inconspicuous, the greenish flowers ultimately fade as the berries begin to develop. At first they are green, but they pass through a series of colour changes – yellow and orange – until they are bright scarlet in the winter. Poisonous to man, other animals which come to eat them will help with seed dispersal. Although the plant dies down in winter, and might be considered dead, with no visible signs of life appearing above the surface, this is not so. With a supply of food stored in a fleshy root-tuber, the white bryony is undergoing a resting period during the unfavourable winter conditions. Once the warmer spring weather arrives, it will start to grow again.

Because some plants have become closely associated with the hedge, this word appears in their common name. Take jack-by-the-hedge, for example. Also known as garlic mustard it grows commonly along the hedgerow providing that the aspect presents it with a degree of shade. Jack-by-the-hedge is seldom encountered as a solitary species. Where the plant become established large numbers seem to thrive in colonies which may number hundreds.

The leaves, heart-shaped and with a toothed margin, have a distinctly garlic smell when crushed between thumb and forefinger, hence one of the plant's common names. The flowers which appear in spring are small and white. The petals are arranged in a star-like formation, and are one of the hedgerow's earliest species. It belongs to the same group of plants as the common garden wallflower. It attracts the attention of the female orange tip butterfly. She lays her eggs here, and once they hatch the garlic mustard provides the caterpillars with a supply of food.

After the flowers have been fertilised they begin to fade. By autumn an abundance of seeds will be protected inside cylindrical seed cases until they are ripe and ready to be liberated.

In days past, if less so now, the plant was gathered by our ancestors because it had specific uses in the kitchen. They relished the flavour which the leaves gave up, and it was used primarily for making sauces, especially to garnish mutton and salt fish. For this reason it was often known as 'sauce alone'. Like so many other common plants of the hedgerow it had its medical uses. Gerard was aware of its properties when he told us that if the seeds are bruised and then boiled in wine the liquid could be drunk to cure several complaints, including colic, wind and stone. Other people discovered yet another property. Apparently anyone who was suffering from hysterics could get rid of the condition by sniffing the bruised leaves.

Of all the early spring hedgerow flowers none is surely more beautiful than the celandine. With its burnished golden flowers and its glossy heart-shaped leaves, no wonder earlier poets have waxed lyrical about its beauty. Wordsworth extolled its virtues when he said:

There is a flower, the lesser celandine
That shrinks, like many more, from cold and rain.
And the first moment that the sun may shine
Bright as the sun itself 'tis out again!

The word 'celandine' comes from a Greek word which means 'a swallow', a reference to the fact that the flower is usually at its best when the swallow returns to our shores.

Containing Vitamin C, the leaves were collected from the plant and eaten, often in early salads as a preventive for scurvy. Alas, all too soon the flowers fade and die, but beneath the surface of the hedgerow the tubers remain, resting, ready to spring forth next year, and repeat their attractive springtime performance.

Soil lives!

At the base of the hedge, hidden and protected under the soil there lives a myriad of creatures. Some of these are so small that it is impossible to see them except with the magnifying power of a microscope. Others, less microscopic, may occasionally be seen if the soil is disturbed. Although we know little of the life of the animals which live there, nevertheless many, if not all, are important to the well-being of the soil, and so to the plants which grow there.

Everyone is familiar with the earthworm, although there are many different species inhabiting the soil. What of the other creatures? Some, like the centipedes, have been mentioned elsewhere. Fast-moving, these carnivorous invertebrates will find a good supply of food in the soil. More leisurely in their ambulatory saunders, millipedes rely on dead and decaying materials.

Slugs, spiders and woodlice keep company with a wide variety of grubs, which, providing they are not eaten by other marauding creatures, will turn into a host of different insects.

A WATERY WORLD

Ponds, lakes, rivers and reservoirs

The variety of water in urban situations is often underestimated. In fact it is surprising just how much water there is in built-up areas, ranging in character from often derelict dockland sites to seemingly polluted canals, and taking in cleaner lakes, ponds and reservoirs. One area not seeming to contribute a great deal is the garden pond. Collectively they can be a real asset, especially in providing a sanctuary for frogs and toads, and an important breeding site for these aquatic animals.

Dockland water often covers larger expanses than most people realise. One of the London dock systems, for example, has the largest area of enclosed water to be found anywhere world-wide. Within the perimeter fence, in areas where the docks are falling into disrepair, there is a veritable haven for wildlife where both plant and animal life will remain something of a mystery because of the security which persists in such areas.

Canals creep into the heart of many conurbations, providing a refuge for wildlife. Aquatic plants in particular are likely to find a niche here. The species which occur will depend on the nature of a particular stretch of canal. Whether it is disused or actively worked will have a bearing on not only the vegetation but the animal life as well.

By far the largest expanses of water to occur in urban situations are the reservoirs. With the continuing need for freshwater for domestic and commercial purposes, they have been built near, or even within, town boundaries. The success of one species of bird in particular has been the result of the increase in open

areas of water which the reservoir provides. This is the great crested grebe (*Podiceps cristatus*).

One might not expect sewage farms to be a valuable habitat for wildlife, but they are extremely important in many areas. Because of the nutrients in the effluent, algae and protozoa flourish. These provide much valuable food for small worms, which in turn attract the attention of various groups of birds.

Thus there is a diversity of aquatic habitats in towns, clean and less clean, large and small, managed and unmanaged, all providing important habitats for wildlife.

Birds

There are some animals, the amphibians like frogs, toads and newts, which rely on water for part of their life cycle. Other species, like some of the water birds, while perhaps surviving away from water, are very dependent on it. Into this group of birds fall the many species of ducks, the moorhen and coot. Other birds like the kingfisher, grebe and heron have need for aquatic haunts when searching for their food. Some birds have particular preferences; others are not as fussy about the type of water.

PIED WAGTAIL (*Motacilla alba*)
Pied wagtails have taken advantage of towns. Once they built their nests among rocky streams, now they will also build in holes in walls.

In one car park, a pair decided to build under the mudguard of a car. They had to suspend their operation when the car went out at night, but this did not, apparently, deter them.

They will nest in large numbers in some towns. Perhaps the most famous roost was of 3000 birds which took a liking to O'Connell Street in Dublin, returning nightly to settle down!

PIED WAGTAIL
The pied wagtail has found a niche in almost every habitat, and seems to establish itself near to man. Often close to water, the bird will also move into the centre of cities. It discovers a wide variety of haunts for its nesting sites, varying from buildings to pollarded trees.

Increases in grebe population

One of the successes as far as reservoirs are concerned centres around the great crested grebe. At the turn of the century, the population of grebes was decreasing rapidly. The fault was firmly at the door of man. As ladies' fashion dictated that the feathers of these innocent birds were needed for decorating feminine headgear, the bird was soon in danger of becoming extinct. Fashions changed, more reservoirs appeared, and the great crested grebe increased its numbers. Fortunately its numbers are now fairly stable, as it is a protected bird. As other areas of inland water have appeared, even more sites have been made available for this once-rare species.

GREAT CRESTED GREBE
Participating in elaborate courtship rituals before mating takes place, great crested grebes offer each other various tit-bits as part of the ceremony.

LITTLE GREBE (*Podiceps ruficollis*)
The little grebe, or dabchick, is a small bird, and very secretive by nature. As soon as it senses any sort of danger, however remote, it will dive, making its way under water to the nearest plants which will conceal it. It will seldom be found far from vegetation, which will afford protection in its almost continuous emergencies.

Cautious, and wishing to discover whether danger has passed, the bird will keep most of the body submerged, simply lifting the head and the neck above the surface.

After the young hatch they are able to swim. When the birds utter the alarm note of 'whit-whit', the young will climb onto their backs.

The courtship ritual undertaken by the birds before they mate is one of the most fascinating of any British species. There are a number of stages. The birds will circle around each other, and touch beaks. This is often followed by a 'peace offering', when each of the birds catch a fish, and hand it to their proposed partner, or, instead of fish, weed may be offered.

Once diplomatic relations have been established, and each bird has approved the other as a mate, nest-building will take place. Both cock and hen bird will set to, and in some cases the nest may be a floating structure; in other cases it may be anchored. The female will lay from three to five eggs. Although they start off with sparkling white shells, it is not very long before they become stained by the water weed on which they are laid. Both cock and hen bird will assist with incubation. After twenty-eight days the young break out of the shell in which they have been protected and nurtured. Once out they will be fed until they are ready to leave to live their own independent lives within nine or ten weeks. Food varies, and consists of both water plants and animals.

Ubiquitous mallard
Although the great crested grebe may only be seen in some urban environments there is one water bird which has become very much a part of the scene in built-up areas. This is the mallard (*Anas platyrhynchos*) which, provided water is available, seems to thrive well

GREAT CRESTED GREBE
Large areas of water attract the great crested grebe, increasingly so as the rate of construction of reservoirs has accelerated with the need for more and more water. The birds are frequently encountered on open water, a habit which is foreign to some other water birds.

MALLARD

Almost every area of water will interest the ubiquitous mallard, and no one will challenge the title given to it—'the best known British wild duck'. Because it is a surface feeder it is able to survive in waters which would prove unsuitable for other species.

where man is found. Not needing the vast expanses which grebes seem to require, mallards will be found on town ponds and lakes, even in the very heart of towns and cities. The mallard is probably the best known of our British ducks. Because it is so familiar, its interesting courtship ritual is probably ignored, except perhaps by the interested ornithologist.

In the breeding season there can be no confusion between duck and drake mallards: they are easily distinguished. The male is the colourful one – with his almost shiny-green head feathers, purplish-brown breast and white collar, he makes the female look particularly drab. She is brown, and the drake takes on an almost identical shade in autumn. He can then be distinguished from his mate, especially when mixed groups appear, because he is darker.

In autumn a number of drakes will start to lay claim to a mate. In an aerial display they will chase and circle her. Later they will again crowd round her on the water, almost trapping her, until mates are finally selected.

The breeding season may begin as early as February. The well-concealed nest is built from a variety of leaves and grass, which will be made comfortable with a lining of down. Although most nests are on the

SHELDUCK

Once driven from rivers because of pollution, the shelduck gradually reappeared as pollution decreased. It has made a comeback along parts of the cleaned-up River Thames.

SMEW (*Mergus albellus*)
This winter visitor to England has managed to make its way along our waterways, and has been discovered well up some rivers, including the Thames in London. An attractive duck, it has always drawn the attention of writers. Its black and white plumage has given rise to a common name of magpie diver in some parts of the country.

A saw-bill duck, the edges of its bill are serrated, and this adaptation helps the bird to grip its food—mainly small fish—more easily.

The first smews usually arrive for the winter in November, and most will have gone by March. The bird does not breed here, and will not be seen except during its winter stay.

TUFTED DUCK (*Aythya fuligula*)
A diving duck, this species had not bred in Britain at the beginning of this century. It has now done so well that it is our most common diving duck. However, although it is quite a common breeding species in London, its status outside the capital is more precarious, and it does not breed as readily.

The bird was first discovered breeding in northern Britain in 1849. Now the breeding population is thought to be in excess of 5000 pairs.

Birds in local parks tend to become tame during the summer. Numbers increase greatly as wild migrants arrive. As many as 500 birds have been recorded on the water in Kensington Gardens. Less than 6km (4 miles) from the centre of London over 2000 birds spent some of the winter on the reservoir at Stoke Newington.

TUFTED DUCK
The tufted duck will dive to a depth of 3m (10ft) in order to find food, which consists of both aquatic invertebrates and plants. As long as there is water deep enough for it to dive into for feeding it may settle in and breed. It occurs on city lakes where it breeds quite happily.

ground, some mallards nest in trees. After four weeks, the young will hatch, the duck having been solely responsible for their incubation. Well developed when born, they will soon leave the nest, and some forty-five days after hatching they are capable of leading an independent existence.

Ways of feeding
Depending on the conditions which are available many species of duck may have taken to a particular town situation. Feeding-wise, it is possible to divide the ducks into four main groups: surface feeders (or dabbling ducks), freshwater diving ducks, sea ducks and sawbills. The most frequently encountered in the majority of towns are those from the first two groups.

Surface feeders take their food from the surface or just below the surface of the water. It is this group which contains some of the most attractive species as far as plumage colour is concerned. Amongst those to be seen in many urban situations are the teal (*Anas crecca*), mallard, widgeon (*Anas penelope*), tufted duck (*Aythya fuligula*), mandarin duck (*Aix galericulata*), pintail (*Anas acuta*) and shoveler (*Anas clypeata*).

The shoveler is aptly named, and, although the beak is not actually shovel sized, for a duck it is large, and the bird cannot be confused with any of the other species which might be found. A useful evolutionary development for surface feeders, the bill has become widened to act as a sieve so that water effectively drains out and food material is retained. The shoveler eats both plant and animal matter. Leaves, seeds and the buds of water plants are taken as well as small freshwater animals.

Both duck and drake have large bills. The duck is brown with speckled underparts. In breeding plumage the drake has a dark green head, chestnut flanks and a white breast. When in eclipse plumage he is much like the female, although the feathers are darker.

The nest is a hollow in the ground, and this is lined with vegetation – mainly grass. A further layer of down is added, and the eggs – laid in April – are placed on this. Between eight and twelve in a clutch, they hatch after twenty-four days.

Britain's smallest native duck
The teal is another species found in the urban situation. It has the distinction of being Britain's smallest native breeding duck. Originally a bird of wooded areas, the

RED CRESTED POCHARD
The red crested pochard is found mainly on larger expanses of water, especially where waterside vegetation, such as reeds, provides good cover. Its appearance has been limited to the south of England.

teal has come to accept the town environment, which has allowed it to expand its territory or provided it with an alternative habitat, especially in parts of the country where woods have disappeared. In the wood the bird had to develop a means of moving quickly when danger threatened. Would-be woodland predators are not easily seen, and as a means of escape the duck rises almost vertically out of the undergrowth, flying into the air like a stone out of a catapult.

Both duck and drake have green patches on the wing. The rest of the duck's plumage is mainly brown. The drake is grey above, and the chestnut head is enhanced and highlighted by a stripe of almost metallic green around the eyes, which continues to the nape of the neck. Above each wing there is also a distinctive white stripe.

A hollow, situated in amongst vegetation, is lined with a collection of leaves and bracken, where available. A soft layer of down is placed on top of this. Eggs are laid in April and May. They hatch twenty-one days later.

The kingfisher's haunts
Provided conditions are favourable, that living jewel, the kingfisher (*Alcedo atthis*), may take up residence along banks of streams and canals. Even if they do not live there, they may come to the area in search of food.

Once a suitable site has been found, the birds will excavate a tunnel in the bank. This completed, they will widen the far end to make a nesting area. After lining it with regurgitated fish bones, the female will lay her eggs. The nest site is generally conspicuous because of a line of lime from the entrance to the tunnel down the bank. This comes from the fish bones which are disgorged by the birds.

We admire the bright colours of the kingfisher's plumage – its upper blue-green feathers giving off a distinctive sheen – but it is there for a purpose. Other animals have come to realise that these feathers are a warning to stay away, because the bird has particularly obnoxious-tasting flesh.

Stately heron
Graceful and elegant, the common heron (*Ardea cinerea*) is perhaps one of the most attractive of the water birds, after the kingfisher. Making its appearance around reservoirs, the apparently sedate elegance is a betrayal of its true nature, for when in need of food its viciously-shaped bill will soon impale unsuspecting fish. Although the feathers are not exactly drab, they blend in well with the waterside vegetation. Although fewer in number than other species, the birds have a wide range, being found over much of the British Isles. As long as there is a supply of food, the heron may be found. Food in particular includes fish, but the bird will also relish frogs, and may take a vole, a tasty morsel.

Frequently seen in towns and cities, even in densely built-up areas, herons are not always popular with those gardeners who have fish ponds. They will continually raid these once they have located them as a

suitable supply of food. In addition they will travel over large areas in search of food. Their tactics here usually vary according to the type of water. As far as the garden pond is concerned, the bird may swoop down, whereas in a lake or along the edges of a reservoir, more patience may be needed. Standing immobile in the water, with neck outstretched, it will ever be on the lookout for food. Once a fish has been spotted, the axe-like bill is down on it in the twinkling of an eye. Sometimes the bird may move back to the bank to consume the fish, but it may swallow it where it stands. Stealth is the keynote of the heron's food-hunting expeditions. If it does not stand still while searching for its food, it will move slowly, noiselessly and deliberately through the water, until it spots something suitable.

There are several species of bird which are unusual in that they can produce their own 'scouring powder'. The heron is one of these. These powder feathers grow all the while, and then break down to form a fine powder. The bird uses this to remove all the undesirable matter – mainly plant material – which the plumage inevitably picks up whilst the bird is on its food hunting rounds. It spreads the powder with its bill.

Herons seldom nest on their own, but are usually found in colonies, called heronries. Typically tree tops are favoured for nesting sites.

Continuous food supply

Within most built-up areas one unmistakable bird will be found, the mute swan (*Cygnus olor*): 'mute' because they are supposed not to be able to utter sounds. Feeding mainly on aquatic vegetation, the birds will usually find as much as they need on most large ponds

PINTAIL

A glance at the bird's needle-like tail will soon reveal how the pintail came by its name. Although pintails are found throughout the year, only a few will nest. Other birds will come to join the resident population during the winter, reaching these islands from as far north as Iceland.

and on lakes, as well as along rivers. As they swim about they suddenly thrust the head and neck into the water. Here they will browse on the submerged vegetation. As with many species which live in our midst the mute swan has come to realise that the urban-dweller will also provide it with food. Although they

GREY WAGTAIL (*Motacilla cinerea*)
Although the grey wagtail is mainly a winter visitor to towns it has managed to find a niche in some, and here it breeds. Although the 'grey' in its name may suggest a drab bird, this is far from the case. In fact the bird has attractive yellow feathers on its underside.

Typically one would not expect the bird to occur in built-up areas, because it is generally fond of quickly moving streams, except in winter when it moves to calmer spots. Nevertheless, because its natural habitat is diminishing, it has found sites in built-up areas.

The female selects a position close to water, and for preference will choose a hole. Here she constructs a nest using a variety of grasses and moss. She will then lay five eggs anytime from April to June.

never seem to become tame, mute swans will readily take food from the bank.

Mute swans can be seen on stretches of rivers within cities, and those on the Thames in London, from London Bridge, and the Avon at Stratford, spring rapidly to mind. The swans swimming unfettered on London's river have owners. Those which do not belong to HM the Queen are the responsibility of either the Vintners Company or the Dyers Company. Annually a well-known ancient ceremony, known as swan-upping, takes place. To distinguish the swans belonging to Her Majesty from those which belong to the two companies, the former have their bills clipped (nicked) in the annual ceremony.

Coots and moorhens

Apart from the mallard, an almost universal inhabitant of the water within and close to towns and cities, the moorhen (*Gallinula chloropus*) is encountered almost as often. Closely associated with this bird is the coot (*Fulica atra*), and the two are often confused, although there is no reason why this should be. The coot is the one with the white beak and shield separating the eyes; the moorhen is red on the beak and between the eyes.

Both species are generally vulnerable on most park lakes since there is very little cover. In other aquatic habitats they rely on rushes and reeds, not only to provide shelter in the event of impending danger, but also to provide secluded sites for their nesting activities.

KINGFISHER
The kingfisher has been aptly described as a 'living jewel'. Occurring along waterways, it will not be found where there is pollution. Although it frequents rivers and streams, it also makes its home and rears its young along the banks of canals and reservoirs.

Both birds are now associated with the urban environment, but this has not always been so. It is only in recent times that there has been a population explosion, and with it the exploitation of built-up areas to provide extra living space. The coot first moved to these areas in the 1920s; the moorhen towards the latter half of the nineteenth century.

Moorhens seem to be able to adapt better to the town situation when nesting. Coming onto land to nest, although this is usually not far from the water, they will find plenty of places to build. The coot is more particular, needing vegetation in which to conceal its nest.

Insects

Nothing can be more beautiful than the almost transparent and yet iridescent wings of the dragonflies. In suitable conditions, the insect will frequent many

HERON
The grey or common heron may be seen from time to time, especially when it comes to raid the local goldfish pond. Although not a particularly numerous species, the bird has a wide distribution, and because of its large size is seldom mistaken for any other bird. Stately and elegant, the heron stands on guard in its nest high in the treetops (right). Nests are built together in tree tops in what have come to be known as heronries. The only known urban heronry is in the centre of London.

aquatic habitats. Water is necessary because it is here that the female will need to lay her eggs, and where the young larvae will develop until they are ready to emerge and become transformed from a drab aquatic form into a dazzling adult insect.

Because the insects are to be seen on the wing from April through to September, the general impression is that the adults live for a long time, but nothing could be further from the truth. In essence their life may be but momentary. Emerging as the brilliant insect, they will mate, lay their eggs and perhaps be dead within a month.

In the British Isles there are around forty different species of dragonfly, and many will emerge as adults at various times during the April–September period. All females will lay their eggs in water, but the means and the modes they employ vary from species to species. Some will pierce the leaves or stems of water plants, and the eggs are then laid there. Others are much less conscientious, simply depositing them in the water.

The actual mating act is rather more complicated than seems necessary – male and female may remain locked together until the eggs are laid. In certain species, the male, still clasping the female, is taken below the surface of the water, where she goes to lay her eggs. One would expect the wings to be damaged by the water, but this does not happen, because a

MUTE SWAN
The large mute swan is no match for humans when protecting its young. Afraid that its offspring—or even eggs—are in danger, the adults will see off intruders, and the powerful wing has been known to break a human arm.

bubble of air covers these delicate structures. When the female has finished her egg-laying she releases her grip on the object she has been holding, and the air bubble then carries them both back to the surface.

Distinctive territory

When considering the territorial claims of animals, we normally turn our attentions to larger specimens, like birds and mammals; but in the insect world, some male dragonflies mark out jealously guarded territories. Although the male will allow any females entry into his domain, he will viciously attack any male having the audacity to enter it. Many adult males show signs of their battles, with tattered and torn wings, although such injuries could also have been caused by birds trying to catch the dragonflies.

The aquatic larvae, which appear to be lethargic, are not particularly attractive or beautiful creatures. They crawl around in the mud at the bottom of ponds, lakes, and streams. Because the body is covered with a coating of hairs, much mud and debris becomes

BIRDS
Many species of birds turn up from time to time on urban waterways, including the larger rivers, like the Thames. A razorbill found it difficult on one occasion to negotiate Westminster Bridge. The guillemot, gannet and grey phalarope have also been recorded.

Pochard have been seen as far up the Thames as Westminster, but the greatest sight for many bird-watchers was a flock of several thousand at Woolwich.

COOT (above)
Coots are noisy, quarrelsome birds. They often gather in large numbers, and then try to chase each other away, as they seek out mates. The coot can be distinguished from the moorhen because it has a white face shield, whereas the moorhen's is red.

MOORHEN (below)
The moorhen is less of an exhibitionist than the coot, preferring the cover which waterside vegetation provides. However, in areas where they live close to man, they have become aware of the benefits, and will come for their tit-bits just as other species do.

attached. The dragonfly larva is capable of fending for itself, and catching food. Particularly aggressive, like the adult it is wholly carnivorous.

The life spent in water varies from species to species. In some it may be years; in others months. Making the transition from an aquatic to a terrestrial form is an important development in the life of the dragonfly larva. Cautiously it will make its way up the stem of a suitable water plant, the movement being slow and laborious. Once it has totally emerged from the water, the nymph's skin splits and the adult will extricate itself. At this stage it is particularly vulnerable to any predators which might be around, but it is unable to do anything about it.

The soft body and wings need to set before the fully fledged insect is capable of flight. Slowly, almost imperceptibly, the fragile wings fill with liquid to expand and give the insect the power of flight.

Under surface dwellers
Other insect species dwell below the surface of the water in the murky depths, and they too must make the transition and become winged creatures. One of these is the mayfly (Ephemeroptera). The time of emergence varies from species to species, but many make their escape in May.

Mayfly nymphs have three prongs to their bodies. Some nymphs' forelegs are adapted in such a way that the insect is able to use them effectively for digging a burrow. Other species, although also living on the bottom, do not burrow, but spend their time crawling about; there are yet others which are capable of fast swimming movements. Mayfly nymphs must feed, and they take small organisms in the water. In their turn they also provide an important source of food for many other aquatic dwellers, carnivorous insects as well as fish.

Ready to change from the nymphal stage to an adult form, they must make their way to the surface. Some species swim; others are lifted by air bubbles trapped under their bodies. Unlike the dragonfly, they have one intermediate stage between the nymph and the adult. As the abdomen increases in size it will finally split, and the mayfly dun will emerge. There is usually a mass exodus of mayfly nymphs, and where they inhabit the water surface it is soon covered, not only with the dun, but with the larval skin, which they have cast off. As they remain on the surface of the water they will provide plenty of food for various animals, and trout are particularly partial to them.

Once ready to fly they will make a short but nevertheless important flight to some nearby vege-tation. Here they are able to hide until the final transformation from dun to spinner takes place.

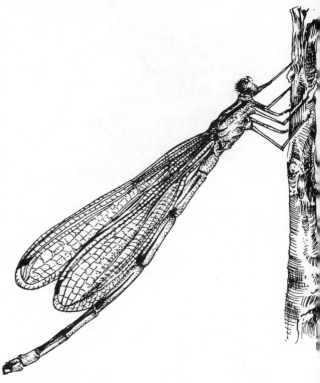

Despite all the trouble in making the change, and in the process managing to survive the ravages of many predators, the life of the adult, though important, is inescapably short. With only a few hours to live its foremost, indeed only, aim is to find a mate, to ensure that the next generation, perhaps spending two years in the water, will replace it when it dies.

On a warm summer evening, the almost leisurely rhythmic flight of hordes of swarming male mayflies belies the fact that there is an important job to be completed before dusk comes, and perhaps death ensues. As soon as a female enters the dancing mob, she will attract the attention of the males, only one being able to succeed in catching her. They fall to the ground, where mating occurs, and the female will then lay her eggs. The male, exhausted by his activities, dies, and once the female has finished depositing her eggs in the water she pauses for a moment, before she too expires.

Case-building caddis

Caddisfly larvae, which as adults bear a superficial resemblance to moths, will also be found in suitable aquatic situations. There are about two hundred different species in the British Isles, and of these in all but one there are aquatic larvae. When the female is ready to lay her eggs, she will find a suitable water source. Some eggs will be laid on water weeds, others under stones. Soft-bodied on emergence, many species of caddisfly larvae will find a suitable form of protection by building cases in which to live. The material used for the case varies from species to species. Some make theirs from water weeds, others cement together

DAMSELFLY
The name damselfly is given to the group of slender-bodied dragonflies. Their flight is much less dramatic than that of the dragonflies, and many species simply flit from one plant to another. Most of the sixteen species found in the British Isles live very close to water, unlike their larger relatives.

sand particles or very small stones, and yet others will use the hollowed-out stems of water plants. Other larvae spin nets in which to catch their food.

Just before a caddisfly larva is ready to pupate, it will take some further protective measures. Those living in cases will seal the end; free-living ones will spin a cocoon around themselves. Eventually the pupa will break out and make its way into the atmosphere where the change to adult will take place.

Movers on the surface

The surface of any pond is covered by a film, caused by the surface tension, which is capable of supporting some of the smaller species which, being light enough, can stand on it. One such species is the water skater or water strider (*Gerris* spp.). Its outspread legs support the light body over a larger area. As its name suggests, the insect skates over the surface of the water, using the middle legs, with a movement like a kind of rowing motion. If the animal does fall below the surface of the water, death could occur by drowning, but most usually float to the surface to continue their movement. The first pair of legs are particularly well adapted to seize its food. Carnivorous by nature, it will feed on any animal which falls onto the surface of the water. It is not unusual to see the relatively small creature tackling a much larger specimen which has drowned – flies are favourites!

Fragile carnivores

Water measurers (*Hydrometra stagnorum*), too, are capable of moving across the surface of the water. Although first impressions indicate a fragile insect, in fact the water measurer is a carnivore, and mainly takes water fleas. Its method of attack is fascinating. Standing on the surface of the water, it is able to pick up

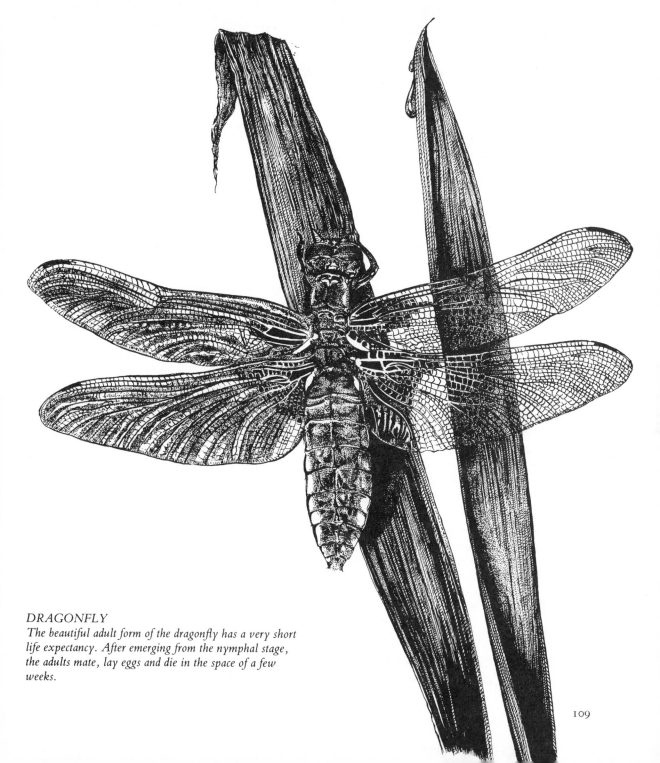

DRAGONFLY
The beautiful adult form of the dragonfly has a very short life expectancy. After emerging from the nymphal stage, the adults mate, lay eggs and die in the space of a few weeks.

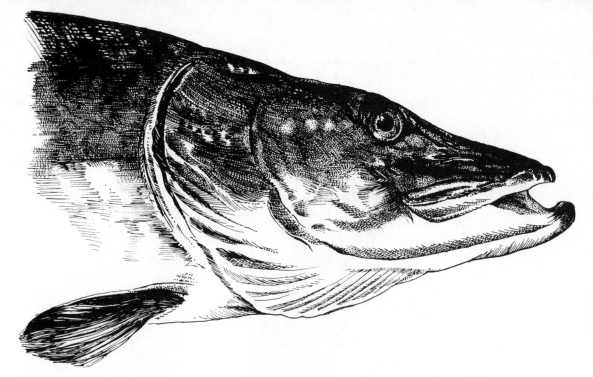

PIKE
Renowned for its ferocious character, the pike uses its mouthful of sharp backward-pointing teeth to catch its prey and push it down its throat. Although the pike can move quickly, it prefers to lay in wait for its food, generally catching slow-moving fish.

vibrations made by the flea. Wishing to make sure that the food is suitable, it will move slowly across the water, and then, with its piercing mouth-parts, seize the unsuspecting animal.

As the colder weather approaches the adults leave the water. They seek out suitable spots on the bank, preferably under stones, and it is here that they will remain until the better weather returns in the following spring, when they will pair. After mating has taken place, the female selects either water plants or stones, and here she lays her eggs. By June the young bugs will have reached adulthood.

Amphibians
Hibernating throughout the winter, frogs and toads make their way to their breeding quarters in spring. What actually stirs them from their deep sleep into a state of almost frenzied activity is something which has, as yet, not been successfully answered. Apart from their need to return to water to breed, these amphibians spend much of their life on land. Indeed, the word 'amphibian' means double life, a reference to their need for both aquatic and terrestrial habitats.

Frogs and toads, although declining rapidly, are found in a wide range of aquatic habitats. Lakes, ponds, slow-flowing streams and canals may all prove acceptable sites, as long as they are free from pollution. Movements of large numbers of frogs occur from time to time, even in urban areas, where there may be so many that they actually stop the traffic on main roads.

1 YELLOW FLAG/IRIS
Once the distinctive iris flowers have been produced and fertilised, the fruits of the yellow flag will develop. They are water-borne, and are therefore light in weight.

2 LESSER REEDMACE/LESSER BULRUSH
Water is a prerequisite for the lesser reedmace or lesser bulrush. It is found around ponds and lakes, as well as beside slow-moving water. Its main means of propagation is by an underground rhizome from which shoots arise.

3 ARROWHEAD
It is the typical arrow-shaped leaves in some arrowhead plants which have given it its name. It also produces leaves of other shapes, and in adverse conditions the arrow-shaped leaves may be absent and the plant will be sterile.

4 FRINGED WATER LILY
The fringed water lily is found in waters which are either slow-moving or stagnant. The flowers which appear in July and August are about 3cm (1½in) across. Although these will produce seeds, which may germinate, propagation is mainly by underground rhizomes.

5 WHITE WATER LILY
Provided that the water is no deeper than about 1.5m (5ft) the white water lily may become established. It may be found in canals, ponds or lakes, where its large leaves float stately on the water, supported by long stalks.

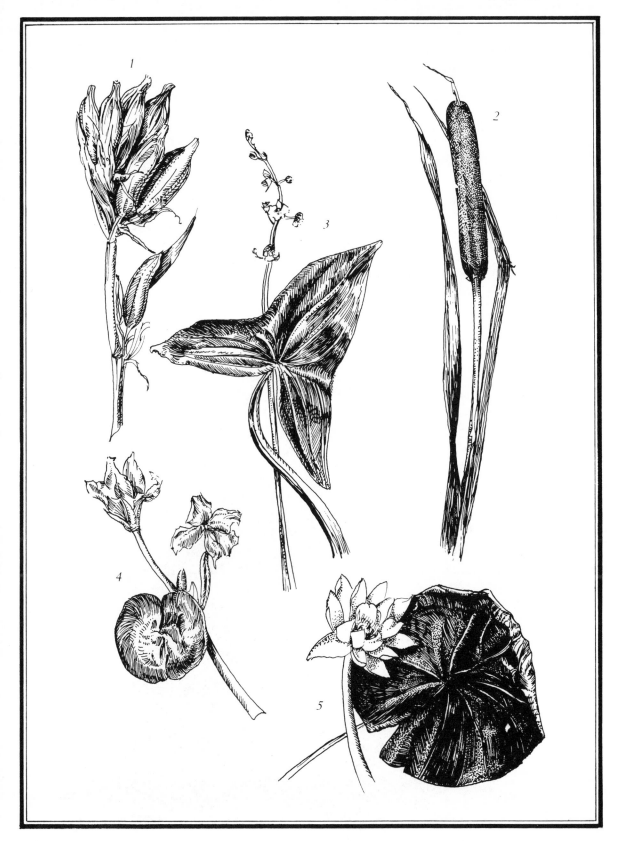

Their need to search for water at the onset of the breeding season far outweighs the over-riding dangers which they face on their journeys.

Although frogs and toads are generally despised, they are exceptionally valuable, because they feed on a wide variety of invertebrates which might otherwise prove harmful to both the farmer and the gardener. Sadly, however, many of the amphibians have disappeared as ponds have been filled in, drainage ditches covered over and other waterways have become grossly polluted. In towns slow-moving streams, once a haven for toads in particular, have succumbed to concrete and now form useless culverts. Poison has wreaked havoc on the population of certain species; collecting too has had its effect. It is, therefore, perhaps not surprising that many naturalists consider that the garden pond is vitally important if British amphibians are not to disappear completely. Suitable aquatic habitats in urban situations will often provide valuable breeding sites.

Capable of camouflage

The skin of the frog is capable of changing colour, so that they are camouflaged. After mating has taken place the female will lay as many as 4,000 eggs. For protection each is covered with jelly. Collectively the eggs are known as spawn. The jelly not only serves a protective function, but also helps the eggs to float, and in addition traps warmth from the sun, so that they will develop. Warmth is a relevant factor which will determine the length of time taken for the eggs to hatch. Generally the tadpoles will emerge after fourteen to twenty-one days. No food is taken for the first few days, the tadpoles surviving on the remains of the

RAINBOW TROUT
Rainbow trout were introduced both for sport and for food, and therefore do not spawn naturally in rivers. They are distinguished from the brown trout by the attractive diffused purple stripe which can be seen along the flank.

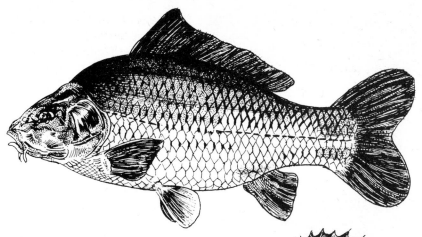

egg yolk. This supply exhausted, they will scrape away at plants, but contrary to observations they are not actually eating the plants but removing the algae which grows there. Some four weeks after their emergence, they lose the external gills. They continue to feed on plants until at the age of six weeks the first signs of the future legs can be seen. As soon as the legs appear there is an immediate change in diet. Vegetable matter is replaced with meat. The tadpole will scavenge any dead aquatic creatures.

COMMON CARP (top)
A powerful body is one of the attributes of the common carp. This, coupled with strong fins and a broad tail, gives the fish a superb turn of speed. It is credited by many anglers with managing to avoid lines because, they insist, it has the ability to learn. Evasive, carp grow to large sizes in waters which provide them with a good supply of food.

PERCH (centre, above)
A prolific breeder, a single female perch may lay several thousands of eggs annually. This may lead to over-population in some waters. A true predator, the perch will take a variety of smaller fish such as gudgeon, bleak and minnows.

BREAM (centre, below)
A particularly adaptable fish, the common bream occurs in almost any water, and manages to master the currents in fairly fast-flowing rivers, just as it does the lower oxygen content in some stagnant sheets of water. Bream seldom seem to be in a hurry and move around, often in shoals, in a leisurely fashion.

TENCH (bottom)
One of the most distinctive features of the tench is the large red eye. Although the casual onlooker might see the tench as a lethargic fish, when danger threatens it moves like a bullet. Most of its speed is generated by its powerful tail.

BANK VOLE

A rodent (gnawing mammal), the bank vole is herbivorous by nature. Although it takes most of its food whilst it is on the ground, the new, succulent leaves on hedgerow shrubs like the hawthorn tempt it to climb into its branches, where it can be seen taking a tasty meal.

Growth continues until the tadpole is three months or so old. At this stage the start of the final change from tadpole to froglet starts. Although it is soon a miniature replica of the adult it will be three years before it is mature and ready to breed.

Warty-skinned toads

It is not too difficult to distinguish the toad (*Bufo bufo*) from the frog. The toad has a rougher skin covered with warts, which produce a vile-tasting liquid, a useful defence mechanism to ward off predators. The frog's skin is smoother, and generally has a mottled appearance. Although the frogs may be found around the water and in moist situations for much of the year, outside its breeding season the toad will be found long distances from any source of water. The toad lays almost twice as many eggs as the frog – 7,000 or so. These eggs are laid in a long chain, which the female wraps securely around the stems of water plants. The life cycle is much like that of the frog.

Plant life

Canals, although they may have flowing water, often present a sluggish character. However, if the water is not too badly polluted, various species of plant may

CORMORANT
Visiting reservoirs, the cormorant will usually find a supply of fish, but it has to leave the water at intervals to dry its feathers out. As they have no oil in their feathers they soon become wet, and once waterlogged the cormorant would drown.

grow there and along the banks. Canals used mainly by holiday craft often have a poor submerged vegetation, but may have many species growing along their margins, including water dropwort (*Oenanthe* spp.) and hemlock water dropwort (*Oenanthe crocata*). Where material from the banks has fallen into the canal, marshy conditions may predominate over small areas, and species capable of colonising such areas move in. Reeds (*Phragmites australis*) and sedges (*Carex* spp.) may manage to establish themselves. Where walls occur along the stretch of canal they will have their own distinctive flora. The variety of plant life varies from one area of water to another, and from one part of the country to another. In some places small areas may support a number of different species, whereas other areas not far away may be almost sterile.

Open areas of water are colonised by plants which can anchor their roots in the substrate (and the depth of the water determines whether plants can do this) and by floating species which do not need to become anchored. The flora around the edges of lakes and ponds, especially where silting occurs, will again be different, and sometimes it is possible to see zonation – a succession of plants from those which grow in the water through to those which are land based, with a gradation in between. Rivers and streams present problems to certain plants, and will exhibit a different flora to open areas of water.

Perennial species
By far the greatest number of aquatic plants are perennials. As autumn approaches they die down and disappear. To overcome the adverse conditions which winter brings, most plants spend this period in the mud at the bottom of the pond, ready to react to the improved weather in the following spring. Some species, like the duckweeds (*Lemna* spp.), make extra food in the autumn. This increases their weight and they sink to the bottom. Once spring arrives, growth begins, the stored food, in the form of starch, is used up, and the plant will float to the surface. Starworts (*Callitriche stagnalis*) which adorn the pond with their star-like flowers will also sink into the mud. Other larger species, including the water lilies, will lose their leaves and stems, but in the mud the rhizomes, filled with stored food, will remain in a state of suspended animation until the following spring, when favourable conditions will once again encourage the plant to grow.

Plants which are typically aquatic have become adapted to the different and exacting conditions which a life in water demands. Plants like the arrowhead (*Sagittaria sagittifolia*) and the water crowfoot have different sorts of leaves, depending on the part of the plant on which they are growing. In the case of the arrowhead, there are no less than three different sorts of leaves. Those below the water are long and thin, and so will not get damaged by fierce currents. Those leaves on the surface are oval in shape, so that they will trap as much sunlight as possible; but it is the other leaves which grow on top of the plant which give the arrowhead its name. These are shaped like the weapons, and point upwards towards the heavens.

The leaves of the crowfoot are similarly adapted, but there are only two forms of leaves. The leaves below the surface are long and thin; those on the surface are round.

Many plants which appear to float on the surface, like the crowfoot and the water lily, are actually rooted in the mud at the bottom of the stream. Some species,

SAND MARTIN (above)

As the boundary between urban and rural communities becomes less distinct, species like the sand martin, which are normally associated with 'rural' areas, may be found. Sandpits on the periphery of built-up areas may provide these birds with nesting sites. It is even possible that, because their original habitats are being threatened, they may be turning to the urban scene for nesting sites. Some have been discovered nesting in drainpipes.

KITTIWAKE (*Rissa tridactyla*)

A member of the gull family, whereas many of its relatives have taken to land in a search for nesting sites and food the kittiwake is a true seagull. It spends much of its time out to sea, and large numbers usually follow boats trawling for fish.

It comes to land rarely, but the breeding season naturally brings it ashore. However, it nests in large colonies along the coast, generally shunning man-made habitats.

Possibly, however, because of its continual increase in population (estimated at about three or four per cent each year), the bird needs new haunts. Cities close to the sea, especially around Tyneside and in Scotland, have 'lured' it slightly inland.

Those which have changed their nesting style have also modified their feeding methods, and like some of their relatives have become scavengers on land.

CANADA GOOSE (below)

Brought to Britain to be exhibited as an ornamental species some two hundred years ago, the birds escaped, and wild breeding pairs are found in many areas. The Canada goose lives where there is grass near water, because it needs to have areas which it can 'graze' regularly.

like the duckweeds and algae, really do float on the surface. Given favourable conditions they will often completely cover the surface of some ponds. Such a blanket cover prevents light from entering the water, and the plants below the surface will suffer, as will some of the animals.

Not all the plants found in aquatic habitats are British. One which has done very well for itself is the Canadian pondweed (*Elodea canadensis*). This foreigner had not been recorded in Britain until 1836. It is thought that some bits of the plant were accidentally brought from Canada with imported logs. Once here, it quickly spread. In less than fifteen years, it had been recorded in all of the canals and on many ponds and lakes as well; but since its initial takeover, when it reached nuisance proportions in some places, it has decreased in quantity, and it now survives along with the species which had already established themselves.

A variety of plants

Those plants which grow around the pond and have their roots anchored in the damp marshy ground include the larger species like the willow (*Salix* spp.) and alder (*Alnus glutinosa*) trees, which will flourish well in damp situations. Smaller plants include the reeds (*Phragmites australis*) and rushes (*Carex* spp.). Bright flowers to look for include the conspicuous yellow kingcup (*Caltha palustris*), also known as the marsh marigold. Especially in days gone by, cattle often grazed the marshes and water meadows where the kingcup grew but they avoided the plant because it contains a poison known as protoanemonine. Although only occurring in small quantities, nevertheless its bitter taste must have proved unpleasant. Shakespeare calls the plant 'mary-bud', and the poet John Gay calls it 'marygold'.

Not unlike the familiar cottage garden forget-me-not, the water plant with the same name will be found in the marshy areas around ponds in particular. A delicate shade of blue, the water forget-me-not (*Myosotis* spp.) has a long flowering period. The first blooms may be encountered in May and there may still be flowers around in September.

Mammals

There are few mammals which live around waterways, but those which do include the water vole (also incorrectly called the water rat), water shrew and of course the otter. In certain areas, particularly the Broadland region of Norfolk and Suffolk, the coypu has taken up residence. Of these mammals, that most likely to be encountered is the water vole (*Arvicola terrestris*).

Although the water vole may not be seen on the land, a slight plop may be heard as it slides into the water. If one is able to detect the direction of the sound, it will not be long before the ever-increasing circles which it sends out as it swims away can be seen.

The fact that the animal is in residence somewhere around the pond or along the bank of a stream or canal is usually given away by the holes it makes in the bank. Usually there are two rows of holes: one almost directly above the other. The top one indicates the winter homes of the animals. They have to move above the water line at this time of the year, otherwise, with a greater flow of water, they would be flooded out. The lower line shows where the animal's summer homes are. The simple hole which leads from the bank may conceal a much more complicated network of underground tunnels. Here there will be feeding and sleeping quarters, and during the breeding season a nursery chamber. Mating takes place from April to October, and there may be three litters a year, each consisting of between three and seven young.

The water vole will feed at given times. During its foraging expeditions it seems almost totally oblivious to other activities around it, but this is not so. Although it is not able to see too well, it does have very good hearing, and should danger threaten it will take whatever action is necessary.

Although it is an aquatic creature in Britain, on the European continent the water vole does not live in water. In Britain it does move away from its normal haunts in search of food.

Swimmers

Several species of fish may be found in urban waters. The bullhead (*Cottus gobio*) generally prefers slow-moving waters, where it will usually conceal itself under stones. It will frequent rivers as well as lakes, and perhaps some smaller ponds, where it will live on the bottom. Dark brown in colour, the body is broken up by a series of marbled and mottled markings, usually of a lighter brown. The belly is much lighter, generally taking on a cream colour.

The female lays her eggs in March or April and places them under stones where the male will take up his sentry duties, his persistent vigil keeping away predators. After four weeks the young will be ready to leave the egg and they reach the maximum length of around 10cm (4in) in four years.

Living by the kill

The pike (*Esox lucius*) is the largest fish to be found in the British Isles, with the heaviest ever caught weighing 24kg (53lb). It is extremely well camouflaged, its mottled green scales blending in well with the underwater vegetation. Pike occur in freshwater ponds and lakes, slow-moving rivers and canals. For preference it will inhabit weedy stretches where it can hide in wait for its prey. A vicious carnivore, its big jaws are equipped with large teeth. So that the fish is able to carry on killing its prey throughout its life the teeth are replaced from time to time.

The fish possesses stereoscopic vision, which allows it accurately to pinpoint and then capture its food. Hiding in amongst the weeds, once its quarry has been sighted, it moves slowly forward, the fins moving almost imperceptibly. Once its prey is lined up, the pike makes a sudden dash of up to twelve metres (thirty-nine feet). Although the adults feed almost exclusively on fish – perch, eels, rudd, roach, bream, gudgeon, trout, etc. – they may take some amphibians and small mammals. According to some people the pike will not kill the tench, which is considered to be the 'doctor' fish. Pike – and other fish as well – are said to rub their wounds against the slimy body of the tench which is supposed to have an antiseptic effect.

Spawning takes place in April or May, the female releasing the sticky eggs into weed. Two to three weeks later, depending on climatic conditions, they will hatch, and can be seen suspended from leaves by means of an adhesive pad. Here they will stay for a week or so. The young feed on small invertebrates in the water, gradually progressing to the fish as they grow. Growth takes place quickly, and they will often be 8cm (3in) within three or four months. Spawning will not take place until two or three years old.

Year-round delights

Viewing the various areas of water will have its rewards at any time of the year. For some species, like birds, the active period is spring and summer; for others it is later in the year. Flowering species are likely to be at their best in summer, but the variety to be found will give an almost all-year-round show of blooms. Insects dependent for survival on the aquatic habitat – the dragonflies, mayflies and caddisflies for example – will be in evidence during the warmer months of the year. Where amphibians breed the spawn of frogs and toads can be encountered during spring, and the developing individuals will be in evidence for several weeks.

Observing wildlife on reservoirs and other open waters will bring its rewards at various times of the year, but by far one of the most interesting sights is the courtship display of the great crested grebe. Other young birds put on a fine show when their parents take them for swimming lessons, and teach them to catch their own food.

Even winter, normally associated with little wildlife activity, has its own charm. Frozen water will prevent birds from feeding but their actions on the ice add a little light relief. At this time they are ever eager for food offered by man. The observer, wanting to discover the activities of wildlife at this time of the year, will have to look for clues. Bird and mammal tracks, oddments of food, droppings – all these will lead to interesting discoveries of the activities of a whole spectrum of life.

The stream

Each aquatic environment is different. To follow the changes in a stream throughout the year gives an indication of what may be seen during the various seasons. Spring in the stream is just as important as spring on the land. Living creatures, which for the most part have been resting for several months, begin a frenzied period of activity, a period in which it is vitally important that mating occurs, so that a particular species will have offspring.

Fish, like those already mentioned, prepare to mate. Others, smaller in size, but nevertheless just as important to the life of the stream, make active preparations. Male sticklebacks exhibit their bright breeding coat, eager to attract a mate. They construct nests to which they entice the females. Once the eggs have been laid, the male ensures that they are protected until they hatch – and his parental care will be exercised for a short time afterwards.

The common minnow lives in some urban waters. It is likely to do well in waters which are favourable to species such as bullheads and trout. Although widespread in Britain, minnows become less common further north. Few, if any, are likely to be found north of the Clyde-Forth basin.

Mating will take place any time from March to June. During the breeding season large numbers of minnows come together, especially where the bottom of its aquatic home is gravelly.

Generally dark in colour for much of the year, males and females are difficult to tell apart, but there is a change of colour in the males in the breeding season. The belly scales take on a distinctive scarlet tinge, and he also sports a black throat.

Each adhesive egg measures about 1.5mm ($\frac{1}{20}$in) in diameter, and will be placed among the stones. In about seven days the young minnows emerge from the eggs: it will be about two years before they are mature. Fully grown minnows measure 10–12cm (4–5in) in length.

Aquatic larvae

As with many other land-living adult flies, the larvae of the alderfly is aquatic. Living and feeding in the stream they eventually emerge as adults. They have come to be called alderflies because their emergence coincides with the emergence of the alder leaves, a tree frequently encountered along the sides of waters, including streams.

Although equipped with what might appear to be strong wings, alderflies spend only short periods on the wing. Once out of the water the adult spends much of its time on vegetation around the banks of the stream. As the alderfly spends so much time resting, it is easy to observe the insect. The wings, folded across the top of the body, give the impression of the ridge on a roof. Having spent many months in the water the terrestrial life of the alderfly is but transitory. With so short a life on land, the alderfly does not even need to feed. Indeed the only activity which is important is mating.

Once this operation is over the female has to get rid of the eggs as quickly as possible. She will release up to 2,000 eggs, which are like miniature brown cigars. These are deposited in groups on plants which overhang the water. Positioning the eggs is very important, as once the larvae hatch they need to fall into the water, because only there can they survive. They are equipped with gills which allow them to breathe beneath the surface of the water. Although they may spend a good deal of their lives crawling around on the bottom, they are able to swim. Using an undulatory movement, they manage to get about quite quickly. Completely different to the non-feeding adult form, the larvae are voracious carnivores, seizing other small aquatic creatures to satisfy their appetites.

Shrimp scavengers

As spring gives way to summer, life in the stream becomes much more numerous and more active. Diatoms develop, and both these and algae provide a much needed supply of food. The amount of debris in the stream will also provide supplies for many of the other creatures living there. One scavenger which spends its life crawling around on the bottom, taking

HERRING GULL
The herring gull is very similar in appearance to the common gull, except that it is larger. More and more, it is nesting on urban buildings inland.

nourishment, is the freshwater shrimp. Smaller males can be found being carried about by the larger females for much of the spring and summer. The initial reaction of most observers is that the invertebrates must be mating, but they are not, and this activity is difficult to explain. The freshwater shrimp has to shed its tough outer skeleton regularly, so that growth can take place. It may do so up to ten times during its life.

Versatile leeches

The rate of flow in the stream determines – to some extent – which species are able to survive. Indeed, survival is only possible for most of the animals which live there because of specific adaptations. Some, like the leeches, manage to survive by hanging on to the underside of stones. Possessing suckers at both ends of their bodies, these are used as the animal moves, which

COMMON TERN
Until recent years the common tern was only found nesting along the coast. Now it has moved further inland, taking advantage of waste-age man.

it does by a series of relaxations and contractions of its muscular body.

Most people still seem to have a distinct fear of leeches. This is undoubtedly due to the 'legends' that they suck blood, the information contained in stories which have been handed down indiscriminately from one generation to another. In Britain there is only one species – the medicinal leech – which feeds on the blood of mammals.

Leeches feed on various aquatic creatures. One species, *Glossophonia complanata*, attaches itself to the body of the water snail, and it feeds by extracting the liquids from the invertebrate's body. The leech's eggs are laid in a see-through cocoon, which may be found attached to either an under water stone or water weeds. The leech needs to be about a year old before it can breed. It mates and lays eggs for the second time when it is about two years old. This is the end of its life cycle, and it dies.

Surface creatures

Not all creatures are found below the water surface. On relatively calm stretches of water, whirligig beetles scuttle about in an almost inexplicable frenzy. They manage to stay above the surface because of the phenomenon known as surface tension. Pond skaters are also supported by this thin skin. Other small aquatic creatures may also be found here. On some wider streams swallows may find an abundant supply of aquatic insects, and hawk for them regularly, their skill enabling them to pick them effectively from the water's surface.

The dying year

Summer gradually gives way to autumn, and the once abundant and active stream is now beginning to slow somewhat. Nevertheless there is still life there. Case-building caddis larvae have been feeding throughout the summer and are now quite large. The material used for the cases varies from species to species. Some use small stones; others pieces of plant material.

Now winter has arrived and life in the stream is much quieter, and even, some might say, sluggish. Of course the creatures which have been there throughout the year to not merely evaporate into thin air. Many spend the winter in the bottom layers, which affords suitable protection. The temperature of the water is much warmer away from the surface, and animals are more likely to survive in these lower layers than nearer to the surface. In most winters, the level of the stream rises considerably with the rains, and possibly the snows; it is much safer away from the turbulent surface. Even the water vole, an inhabitor of the bank, will have to excavate himself some new tunnels above the level of the water, otherwise his burrows will become waterlogged.

Blackfly larvae overwinter in the stream. When they are ready to pupate, a film of air surrounds the body, and this carries the creature to the water's surface. Here it breaks through the surface tension and is ready to fly away.

Although most creatures breed in the spring there are some like the wild trout which use the winter period for this activity. By moving their tail backwards and forwards the female makes a suitable depression in the bottom of the stream. Here she lays her orange eggs which are fertilised by the male. The water temperature will determine how long it is before the eggs hatch. On average it takes about twenty-eight days.

IN QUIET SOLITUDE

Wildlife in the churchyard and at night

There are many areas of the town environment we fail to notice, but probably none more so than the churchyard. One feature of many churchyards is the tall cover of grasses. Not only will this provide plenty of cover for many invertebrates, but it is ideal ground for many small mammals. Grass need not be so very long to be a valuable asset, and grass left at about 8cm (3in) will give adequate cover for many of the small mammals, like shrews and voles. Both the pygmy (*Sorex minutus*) and common shrews (*Sorex araneus*) are likely to occur. These small insectivorous mammals need to feed regularly, eagerly devouring a whole range of invertebrate life which lies hidden in amongst the stems. In order to survive the shrew needs to be feeding constantly. Not only are shrews active by night, as well as day, but even in the harshest winter

they will need to be out and about in search of food.

Shrews are common in urban areas and, if it is hard to see them, it is often easy to hear them. They are quarrelsome animals and one can often hear their high-pitched shrieks as one shrew comes across another. They seem to find plenty of food in hedgerows around churchyards, as well as in amongst the grasses and other vegetation. Even in the densest cover they are not safe; their presence is often discovered by foxes and stoats, and domestic cats kill their fair share.

Grass will provide these small mammals with cover, so that they can construct their runways. Common shrews make a labyrinth of tunnels, but pygmy shrews are not as industrious. They prefer to use those made by their relatives, or by voles. The runways are a means of moving from one area to another under cover.

TAWNY OWL
The tawny owl is predominantly a nocturnal hunter, but there are times when food is scarce. When this happens the bird will come out to seek a supply during the hours of daylight. In many parts of the British Isles it is our most common owl.
IVY
The evergreen ivy is a valuable plant for many birds. When most other plants have shed their fruits and seeds, the ivy produces its berries, deep in the cold winter weather. Earlier, in the autumn, when other flowers have faded, the ivy comes into bloom, and provides a plentiful supply of nutritious nectar for many insects. In their turn, they will ensure that pollination takes place.

The larger of the two shrews is the common species, but even it is quite a tiny animal, measuring no more than 12cm (5in) of which 4cm (1.5in) is tail. The snout is elongated and carries many whiskers. This shrew weighs between 6 and 12g (0.2 and 0.4oz). Above, the rusty brown fur is close and dense; below it is lighter, usually taking on a greyish-yellow tinge.

Most common shrews will mate twice in a season, first in spring and then in summer. A ball-shaped nest of loose woven grass is placed under some protective feature, such as roots, a stone or a log. Within three weeks of mating between five and ten young will be born. Blind at birth, their eyes will open after ten days. The female looks after them for between three and four weeks. Most will be dead within a year. Thus a shrew is likely to see only one breeding season, and many females will die before their young are weaned. They provide food for predators, although some may not find them particularly palatable, because of an evil-smelling/tasting liquid which they give off from musk glands situated on the flanks. Owls find them edible, and therefore they provide valuable food.

When taking its own food, the shrew produces a poison mixed in with its saliva. When this is injected into a larger invertebrate, which the animal might not otherwise be able to tackle, it controls the prey.

Homes for voles

Both field and bank voles are likely to find many churchyards hospitable places. The bank vole (*Clethrionomys glareolus*) will be seen from time to time, but most of its life is spent moving around in tunnels, some above the surface, others below it. It is more likely that you will hear the animal as it makes a rustling sound when moving about in the vegetation.

Whereas shrews are characterised by a long snout, voles have blunt faces and plump bodies. Although both species do occur, you are more likely to come across the field (or short-tailed) vole (*Microtus agrestis*). With fur which varies from grey to russet, the animal has extremely small ears which are almoıt undetectable.

The tunnelling system the vole makes for itself consists of a network of pathways which cross and criss-cross each other, so that it is difficult to imagine the animal getting anywhere! At irregular intervals along this network the vole has its feeding quarters. It busies itself collecting together a store of freshly-cut grass, which it carries to its 'dining room'. Here it apparently feels much safer as it consumes the succulent stems.

The field vole has a long breeding season which may extend from March through to September. Like the shrews, most will only breed for one year: they will be dead before the arrival of the next breeding season. A nest is necessary to house the young mammals when they are born. This can be found either in a chamber below the ground or in a depression on the surface.

With litters varying in size from three to six, and with the young voles reaching maturity when a month old, there would probably soon be a population explosion in any one area if many of the animals – adults and youngsters – did not fall prey to other species. Owls take a large number. Voles also form a valuable part of the food of stoats, weasels, foxes, hawks and snakes.

Underground worker

Another animal which spends much of its life below the surface of the soil is the mole (*Talpa europaea*). Although the animal was originally found only in grassland and woodland, as these habitats have decreased the animal has had to make its home in other areas.

Living in darkness in its subterranean world, the mole does not need to see; but its other senses are necessary and well-developed. Both smell and hearing are of great importance to the creature. Vibrations passing through the soil will convey a variety of information: these vibrations will be picked up by whiskers which are found on both the head and the elongated snout.

Torpedo-shaped and sleekly built, the mole can move through the soil easily. A further adaptation in the form of its spade-like forelimbs aids its efforts at forward propulsion and tunnelling. It is only the forelimbs which are strong; the hind limbs are very weak. The mole loosens the soil with the front feet, and throws it backwards.

Sadly most of the mole's activities cannot be observed by man. However, it is considered an extremely active and hard-working mammal. It has a regular cycle of activities. Working almost solidly for a four-and-a-half hour period, it will then rest for the next three-and-a-half hours in order to recover some of its expended energy.

It is only at the breeding season that the solitary-natured boars and sows take an interest in each other. This is purely a temporary phase, and once they have mated the female will be concerned with her future confinement; the male with his survival. Litters vary in size between two and seven offspring. Without fur, and blind, their well-being is in the hands of the female alone. Just over one month after the helpless creatures came into the world they no longer have any need of their mother, and will leave her to seek their own food. They will be ready to pair off when the following spring comes.

Nesting sites in trees

Trees feature in the flora of many churchyards. A lot of these are old, and so provide suitable nesting sites for various birds. With a good supply of mammals to feed on, both tawny (*Strix aluco*) and little (*Athene noctua*) owls may be found. If they do not actually live there it is likely that, particularly on the edge of built-up areas, they will come to visit. Owls, with their silent flight, are characteristic of churchyards, and have often given rise to the stories of ghosts which abounded in days gone by.

The woodpeckers

Once trees have become old they also tend to become unsafe, and have to be removed even from church-yards. However, not all old trees fall into the unsafe category, and it would be much better if some of these could be left. Besides providing nesting sites for the owls, they may attract woodpeckers. Species which may be found in churchyards are the lesser spotted (*Dendrocopos minor*), great spotted (*Dendrocopos major*) and green woodpecker (*Picus viridis*). Although they are generally found only in suitable places on the edge of towns and cities, the great spotted and green species sometimes manage to penetrate further into built-up areas.

Although the most common species in towns at the turn of this century was the lesser spotted woodpecker, it has now declined considerably. Many people have speculated on the reason for this. Some have suggested that its decline has been due to the expansion of the area needed by its larger relative the great spotted woodpecker. Early in the 1800s this bird was rapidly decreasing in numbers, and yet a hundred years later it was on the increase. It now has the distinction of being the most common woodpecker to be found in the British Isles.

Although trees are necessary for all woodpeckers, and most species will spend quite a long time searching the bark for wood-boring insects, the green woodpecker has learnt to adapt well to living in the urban environment. Apart from feeding on larvae living in wood, it will also come to the ground to take ants and other invertebrates, including moths. Where the opportunity arises, the woodpecker will find some of its food in the garden, especially seeking nourishment from the well-laden bird table. Indeed it has come to realise that other food can be as nourishing as the wood-boring invertebrates. In many areas it has taken the liberty of raiding the nut bags put out for tits. In some gardens it has adapted so well to this that it is now able to copy tit antics, and hang upside down on the bag as it takes its fill. Vegetable matter, like fruit and grain, may also be eaten if the alternative food source becomes scarce. Although the great spotted woodpecker resembles the green in habits, it comes to the ground much less frequently.

The green woodpecker has the distinction of being the largest of the British woodpeckers; it is also by far the most colourful. When a pair find a suitable tree they will both set about the task of digging out a

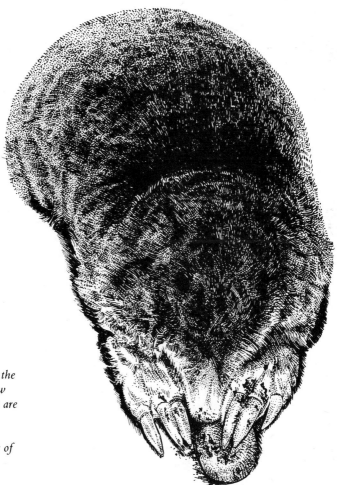

MOLE
The mole is encountered mainly on the edges of towns, where it may burrow under lawns. Although its activities are not always appreciated by the gardener, it does a valuable service when it mixes up the various layers of soil. Its many tunnels also help to aerate the soil as it passes through.

nesting hole. As they repeatedly peck at the bark there is soon a collection of chippings on the ground, giving away their industrious activity above.

Once the nest-hole is complete, the female will lay between five and seven eggs. They are generally laid in April or May, and for nineteen days both birds will share the task of incubating the eggs. Once the young have hatched, both birds continue their parental tasks, as they bring a continuous and increasing supply of food to the demanding nestlings. Three weeks or so after they hatched the young are ready to fend for themselves.

The fully fledged green woodpeckers will now be ready to search out their own supply of food. In the churchyard there is not likely to be enough for them, and this is where the nearby gardens come in. The young will enjoy the same diet as their parents. With a sticky tongue, the birds are able to pick up their food easily. Perched on the side of a tree the bird holds onto the trunk, using its feet. The tail is important too, being pushed against the tree trunk to support the bird.

Beetles abound in their thousands

With over 3,000 different species of beetle in the British Isles, most churchyards will provide a suitable habitat for some species, whether they are resident or visiting. These vary in size from the large 6cm (2½in) stag beetles (*Lucanus cervus*) to some which are only the size of a full stop. In spite of the large numbers of species it is not difficult to distinguish beetles from other groups of invertebrates. Although a beetle has two pairs of wings, the first of these has evolved with hardened elytra or wing cases. Because these plates cover the abdomen they effectively hide the second pair of wings arranged beneath them. Beetles are not alone in having these hardened elytra but in beetles they come together so that the join forms a line down the centre of the animal's back. Although many beetles have well developed wings, enabling them to fly, there are others in which the wings are too small.

Some of the most common and often seen are the ground beetles. Black in colour, they are frequently encountered as they run across paths in churchyards. Equipped with long legs, they run quickly and actively. This fast movement is a pointer to their mode of life. Predatory, these carnivorous beetles move quickly in order to capture live food. Although many are black, they have a characteristic and beautiful metallic sheen, and none is more beautiful than the violet ground beetle (*Carabus violaceus*). They seek out damp sites in the churchyard, and rotten wood, particularly with leaf mould in and around it, is a favourite haunt. Large stones will provide them with an alternative place to hide.

The best time to observe active ground beetles is at night, although they can be discovered during the day in their diurnal hiding places – under the stones and logs. At night, however, they will be out in force, searching for food. Slugs and other small invertebrates feature regularly in their diet.

Many of these ground beetles only have Latin names, like *Feronia cuprea*. The body of this one usually exhibits a copper-coloured sheen and this obvious recognition feature is emphasised by the antennae. The two segments at the base are distinctly red.

Notiophilus biguttatus is often out and about even in bright sunshine. This small, almost iridescent beetle can be seen scuttling about or simply sunning itself. It is disturbed by the slightest movement, and makes off at an incredible speed for such a small insect. It is soon lost to view as it disappears into some minute crack or cranny, which at first sight appears too small to take its body.

Rotten wood will often provide a much needed supply of food for stag beetles (*Lucanus cervus*). Once the female has deposited her eggs in the wood they will soon hatch. The larvae will then spend their time tunnelling in the wood, where they will spend several years before they are ready to emerge as adults. With man's obsession for cleaning up the countryside and removing or burning decaying timber, the stag beetle has suffered considerably.

It is the 'fearsome' antlers of the male which have given the stag beetle its name. Few people realise that these antlers have developed from the jaws. They will not normally be used, except in defence, when the male attempts to ward off would-be attacks from its own kind. On warm summer evenings the large males may be seen on the wing. They fly more frequently than the females.

The grass roots will also provide a source of food for the well-known wireworm. This is the larva of the click beetle (e.g. *Agriotes obscurus*). The female will lay her eggs in the soil, and when the larvae emerge they will start to feed on the roots of grass. In the countryside they can cause a great deal of damage by feeding on the roots of cereals, especially wheat.

The larvae are tough-skinned and have a distinctly yellow tinge. In fact they are regularly turned up in the garden when the soil is dug. Because the adults are nocturnal – and will only be seen during the day if they are disturbed – they are not as familiar as the larvae. Larvae stay in the soil until they are fully grown and ready to pupate. This may be as long as five years after the eggs were laid.

There are several species, but all have the same basic shape. The antennae have serrated edges, which are not unlike the teeth on a saw. The distinctive feature of the

YEW

Yew trees are frequently encountered in towns, especially in churchyards. Our ancestors believed that because they grew to an old age they would help keep away evil spirits. Leaves were often collected from the tree, and sprinkled on the coffin for the same reason. Attractive yew berries provide an incentive to birds to come and take them, ensuring that the seeds will be dispersed. The flesh on the outside of the berry is soft and harmless; it is the seed inside which is poisonous.

click beetle is its ability to get back onto its feet if it should find itself on its back. By pushing the head and the front end against the rest of the body, the insect is able to jump up into the air. Whilst airborne, it performs a somersault, so that it will land back on its feet. There is always a positive 'click' as the insect does this, and this is how it has come by its common name.

Attacks on grass roots

Grass roots in churchyards are the target for yet another species. The culprit is the leatherjacket, which is familiar to many people, yet few people realise that this is the progeny of the familiar daddy-long-legs or cranefly (Tipulidae, e.g. *Ctenophora atrata*).

Each female will lay around 300 eggs, placing them in the ground during the summer months. Although the leatherjackets are only noticed when they are uncovered during digging or ploughing, they will probably come to the surface on warm, moist evenings. Once at the surface they will start to take their fill from the plant stems where they enter the soil. Although the adult craneflies are active throughout the period late spring to early autumn – about May to October – they become particularly noticeable from about August, when large numbers emerge from the ground. Pupation takes place in the soil, and when it is time for the adult to emerge the pupae make their way to the surface, using a wriggling technique. Although the transition is made at various times, perhaps right through to late October, those which do not emerge remain in the soil, feeding during the winter months and pupating in spring.

Timber destroyer

One species which is particularly associated with the interior of churches is the death watch beetle (*Xestobium rufovillosum*). Because of the amount of timber in these buildings, the death watch beetles have a field day in churches. In earlier times, when much more timber was used in houses, they were also common there. The beetle acquired its name because the knocking sound it makes was heard in the silence when a vigil was being kept over the dying.

To make its banging sound, the male hits its head against the inside of the tunnel where it lived and fed as a larva. According to naturalists the reason for its doing this is to attract the attention of a female. Measuring around 12mm ($\frac{1}{2}$in), this particular woodborer is one of the largest to live in wood.

Having mated, the female will lay her eggs in either April or May, each batch consisting of around 50 eggs. As with most species moisture is necessary, and the death watch beetle larva will only survive in damp timber, which is frequently encountered in old buildings.

Although oak is preferred, the sort of timber in which the eggs are laid will dictate how long it takes for the larva to complete the life cycle. In some cases it may be ten years before the larva is fully grown and ready to pupate. Pupation takes place in the autumn, and within a few weeks the insect will emerge as an adult. It will spend the next few months here, eating its way out of the larval tunnel. In the following spring it leaves its distinctive hole, so characteristic of death watch beetle activity.

Below the surface of most infected timber, the larvae will have done a great deal of damage, and although surface holes will indicate their activities what has gone on underneath is usually disastrous, eventually causing the collapse of timbers.

Plant life

Provided that conditions are favourable and that herbicides have not been used, the churchyard can provide a refuge for a large number of wild plants. The individuals vary depending on conditions such as soil and moisture. Mown grass in churchyards will give rise to a different variety of species from those in churchyards where the grass is allowed to grow.

White deadnettle (*Lamium album*) will grow in grassy areas, although it prefers some shade as opposed to full sunlight. It is one of those species which if cut down during the summer will usually grow up again. In reasonably mild weather it can be found growing until Christmas. Where wild rabbits come into the churchyard they will probably feed on the leaves, and where they do they can cause the plant to die off.

One buttercup which is able to grow well in amongst the taller herbs is the meadow buttercup (*Ranunculus acris*). The leaves are deeply cut, and the bright yellow flowers are borne on much-branched flower stalks. They appear from May to July, and will attract a whole host of insects. In addition, in areas where the herbs tend to be shorter, the creeping buttercup (*Ranunculus repens*) will often be found. It is probably this plant, rather than the other species, which the earlier herbalists referred to as the buttercup.

JACKDAW (*Corvus monedula*)
Although jackdaws do come into built-up areas to breed, they are probably seen more often making sorties in their search for food. For some reason their numbers have increased during the present century, and naturalists have put this down to their ability to feed on other materials, as Britain has changed its way of life.

There are breeding colonies in many cities in Europe, as well as large numbers of other birds which, like their fellow humans, commute into urban areas each day.

Jackdaws have a habit of collecting things, hoarding together a whole host of useless items.

Because the bird has lived close to man for many years, a number of rhymes and sayings have grown up around it. People in Norwich, for example, consider:

When three daws are seen on St Peter's vane together
Then we're sure to have some bad weather.

Its descriptive name is a reference to its means of spreading. Runners can be seen coming out from the bottom of the plant. They will take root at the joints (nodes) and so in this way they form new plants. The plant reaches a height of up to 60cm (24in). Buttercups tend to flourish in churchyards even where animals are grazed, because they produce an unpleasant tasting and poisonous juice, which is unacceptable to the cattle.

It is probably true to say that the most frequently seen tree species in many churchyards is the yew (*Taxus baccata*). Some of these – although one wonders how they survive with hollow trunks – may have been growing for around eight hundred years. Because of its longevity the yew was seen as a life symbol by our ancestors, and so revered was it that it was quite common practice for yew leaves to be spread on a dead person's grave. Not only were the trees planted here, but superstitious landowners would often have one situated close to the house, for it was said to protect the people living there. It was, according to earlier folklore, the 'Lord of the home'.

Yet whilst such beliefs were held in some parts, the tree was also referred to elsewhere as the 'death tree'. Those who dared to sit under its boughs, let alone sleep there, were certain to die for their folly. In church-yards, not only was it planted for the reasons already given, but it was also claimed that it would drive away evil spirits. Left unchecked in the churchyard the tree will grow to a height of between 5m (15ft) and 16m (50ft). Although the trunk is not particularly long, it has a large girth, maybe greater than 15m (50ft) in exceptional cases, although most trees vary between 7 and 15m (25 and 50ft). Covered with red bark, which flakes off irregularly, the trunk always looks bedraggled.

A look at a yew which has been felled shows narrow annual growth rings, which means that the tree is slow-growing, and a yew with a large girth may have been growing for a very long time to attain such a size. Because of its slow growth rate, the timber is particularly dense but also elastic. When wood was used more than it is today, it was said that a post made from yew would remain intact longer.

With male flowers on one tree and female ones on another, cross-pollination is necessary. There are occasional exceptions, and a tree which is predominantly of one sex will sometimes have one branch with flowers of the opposite sex from those found on the rest of the tree. The male flowers, appearing in February and March, may often be overlooked, because they occur on the underside of the branches.

Once fertilised, the flowers will eventually give rise to the well-known red berries. Although the outside is soft and has a sweet taste – an attraction to young children – the internal seed is hard and poisonous. So, contrary to popular belief, the berries themselves are not poisonous, but the seed is. It is not only the seed which has poisonous properties. The leaves, too, contain a poison, which can bring about death, though small quantities are unlikely to have any serious ill effects.

Under cover of night

Without making any effort most of us have seen a few of the creatures which bear the tag 'nocturnal', especially those night-time fliers, the moths; but there are many species which come out under cover of dark to feed because at this time they are less vulnerable, and which escape our attention. Because they do not want to be disturbed during the day, these animals will often frequent the churchyard.

Night-flying mammals

Bats are somewhat 'unorthodox' mammals, the only ones which are capable of proper flight. During the course of evolution, bats, like several other groups of mammals, have developed membranes – sheets of skin – which act as wings, but only bats possess the true power of flight. Bats belong to the order Chiroptera, which means 'hand-wings'. Although the upper arm is short, it is extremely strong. The forearm is long, and is terminated by a short thumb and elongated fingers. The wing membrane, which extends from the shoulder to the wrist, continues between the fingers and stretches from the little finger towards the rear limb, where it is attached to the ankle. In certain species the skin continues across the tail. In flight the membrane, containing large numbers of capillaries, serves a similar function to that of a car radiator, keeping the bat cool.

In Britain the twelve species of bats vary in occurrence from common to extremely rare. In the case of the greater horseshoe bat (*Rhinolophus ferrumequinum*), this is now a protected species under government legislation. Of Britain's bat species undoubtedly the most common to be found in the urban environment is out smallest, the pipistrelle (*Pipistrellus pipistrellus*).

Since its food consists of live insects it is not surprising that the bat catches its food on the wing. Street lighting attracts night-flying insects, and bats have become aware that here is an extremely valuable source of food. Indeed, many bats are reported to have moved into people's lofts to be near their suppers. Like so many other nocturnal creatures the bat has been difficult to observe and study, and they often cause alarm or fear. Perhaps such a reaction is not totally without foundation: gliding about silently, the bat has a nasty habit of swooping low as it searches a regular supply of night-flying insects. Although it is difficult to distinguish one bat from another at night, the pipistrelle's flight path is characterised by a series of rapid jerky movements.

During the day the pipistrelle will make its way to suitable buildings. There are often plenty of these in the town, ranging from the most popular – churches – to derelict buildings and warehouses. Where there are suitable daytime resting sites, large colonies of pipistrelles may be found. When the time arrives for the bat to hibernate for the winter, the large groups encountered for the rest of the year tend to disperse, and smaller groups are usually the order of the day. Buildings, however, are still a favourite haunt.

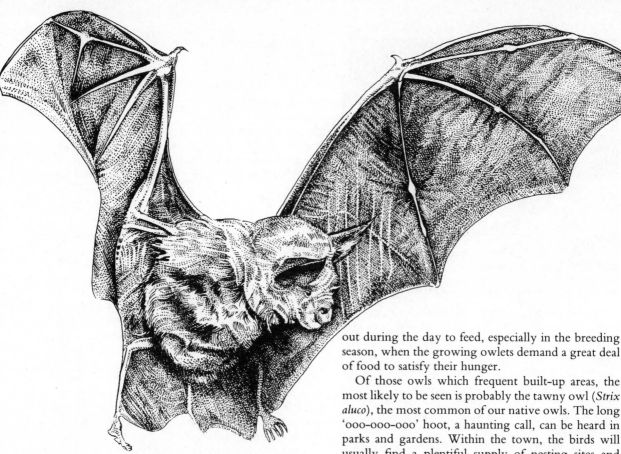

GREATER HORSESHOE BAT
By emitting high-pitched sounds, generally inaudible to the human ear, the greater horseshoe bat (as well as other species) is able to avoid crashing into objects. Each bat has its own particular wavelength, necessary otherwise the sound would be picked up by other bats.

Another bat likely to be found in the night-time townscape is the serotine (*Eptesicus serotinus*). Much larger than the pipistrelle, its flight is less definite, some would say indecisive – in fact, perhaps not unlike that of one of the large moths. During the day, buildings will attract the mammal, and where cracks in walls are large enough these will also be sought out. Although hollow trees prove suitable for hibernatory quarters, other places, like barns, churches, steeples and cellars, are also suitable. Other town-dwelling species include the long-eared (*Plecotus auritus*) and the noctule (*Nyctalus noctula*).

Birds of the night

With a few exceptions, the owls are nocturnal. Living by the kill, they find a good supply of food in the urban situation. Classified as nocturnal, the birds may come out during the day to feed, especially in the breeding season, when the growing owlets demand a great deal of food to satisfy their hunger.

Of those owls which frequent built-up areas, the most likely to be seen is probably the tawny owl (*Strix aluco*), the most common of our native owls. The long 'ooo-ooo-ooo' hoot, a haunting call, can be heard in parks and gardens. Within the town, the birds will usually find a plentiful supply of nesting sites and tawny owls have been recorded right in the centre of towns. Holes in trees are the favourite nesting sites, but the tawny owl is not fussy, and where there are no tree holes it will lay its eggs in buildings. It is not unknown for this species actually to lay their eggs on the ground.

In suitable conditions the female will incubate between two and four eggs, laid either in March or April, in one short breeding season. The owl will quarter many areas of the town where suitable cover conceals small mammals. Churchyards and parks are among the most rewarding areas for the owl on its nocturnal hunting expeditions.

Although one reads that the owl's diet consists mainly of small mammals, it may not find these so easily in the town, and here, ever-adaptable, it turns its attention to a diet consisting almost wholly of sparrows. It will endeavour to vary this diet, and where it can it will include such delicacies as earthworms and large insects, and may even take fish.

Although the bird is, for the most part, nocturnal, if it is disturbed during the day it may be glimpsed fleeing from its diurnal haunts. In town it often seeks high points, such as a roof top or tall tree. Because it is from dawn to dusk that it is on its rounds, it is more likely to be heard than seen.

SEROTINE BAT
The serotine frequents gardens and rubbish tips, its movement characterised by a fluttering flight path. In winter it hibernates in colonies. In the summer its daytime roosting sites will be in similar places.

A newcomer to the British Isles

Although the little owl (*Athene noctua*) is a relative newcomer to the British Isles, having been introduced in the nineteenth century, it has made its way into towns. Of all the owl species, this is the one which is more likely to be seen during the hours of daylight.

Much less common than the tawny owl, some have been known to breed in large city parks, including Battersea in London.

Much smaller than any of the other British owls, an average specimen is not much bigger than a thrush, and measures around 22cm (8½in) in length. Its appearance of having a plump body is somewhat exaggerated by its flat crown. The feathers have a greyish brown tinge, and this base colour is broken up by a variety of spots and bars. It can sometimes be seen in the daytime sitting quietly on a tree or telegraph pole in close proximity to its nest. Disturbed, it will make off with speed, although its flight path tends to take it in an undulating line. Its main food consists of insects, but it takes some small birds and mammals, as well as frogs, reptiles and worms.

Of the owls which may be found in towns, undoubtedly the barn owl (*Tyto alba*) is the most 'appealing'. Yet its distinctive, virtually heart-shaped face has not always been admired; indeed our ancestors were afraid of the bird with its light plumage, a silent flight, and its nocturnal nature. The barn owl has been a regular town dweller for many years. Yet, since the early part of the present century, its numbers have declined. Although it is a bird which is quite happy where there are buildings, it seems to prefer older, generally unoccupied ones – churches, empty for much of the time, old warehouses and so forth – which have to some extent disappeared in the mammoth rebuilding of our cities.

Nocturnal moths

One of the moths frequently found is the garden tiger moth (*Arctica caja*). Although its brilliant markings vary considerably, nevertheless both the caterpillar and the adult moth are likely to be recognised more than many other species. The brightly coloured scales are one of nature's protective rather than deadly features, warning would-be enemies to keep off! Brightly coloured insects often have an unpleasant taste. Others copy this as a method of making other animals keep away. This is known as warning colouration. As with the adult, the caterpillar is particularly conspicuous, with its almost absurd 'woolly' body covering.

In some species of moth the caterpillar is a much more interesting specimen than the adult form. This is true of the puss moth (*Cerura vinula*). The adult is a nondescript mixture of white and grey, but the caterpillar is far from drab. Starting off as a dark larva, almost black in some cases, as it grows it takes on its bright colours. When it acquires its green skin it also sports a red-brown saddle. At the end of the body there are two tail-like projections. Ever on the lookout for predators, it often assumes its threatening position, in which it raises the head and tail. The purpose of such behaviour is obviously to put off any animal which might endeavour to attack it.

Evolved to cope with pollution

The variety of species within any particular group of animals is due to evolutionary changes which have taken place over thousands, perhaps even millions of years, as individual species change to adapt to new conditions. With the coming of the Industrial Revolution, the once light coloured bark of many trees became dark from the effects of smoke. The light coloured peppered moth (*Biston betularia*), at one time inconspicuous, was soon very noticeable on the soot-encrusted buildings and trees. Then, in industrial areas, naturalists noticed that changes were taking place in the peppered moth. Instead of a white insect with black markings on the wings, a black species, formerly only found in small numbers, became predominant. In a relatively short time the black moth grew in numbers, and in some areas more than 90% of all peppered moths were of the black variety.

The peppered moth is not the only species which has adapted to the changing industrial conditions. Entomologists reckon that more than a hundred species have shown some such changes. In some species these have been small; in others extremely obvious. It will be interesting to see what happens as our once-dirty cities and towns are cleaned up.

A moth in winter

Although winter may appear to be an unfavourable time of the year, this is not so for all moths. As its name implies, the winter moth (*Opheroptera brumata*) will be out and about when other less hardy species have retired for the winter. The adults are only on the wing at night. Mating takes place when the male finds a wingless female resting on the trunk of a tree. After they have mated, she will make her way up the tree to lay her eggs. The green caterpillars, called loopers because of their means of movement, which hatch in the following spring will only do well if their emergence coincides with the appearance of the tree's young, tender leaves. Hatching too early means no food; emerging too late will only leave them with older, tougher leaves, which they are unable to eat. It will be late in autumn – probably October or November – before the adults emerge from the cocoons.

Smaller animals seek night-time cover

Two groups of nocturnal invertebrate animals which are frequently confused at first sight are the centipedes and millipedes. Anyone who watches their movements will be able immediately to distinguish between the

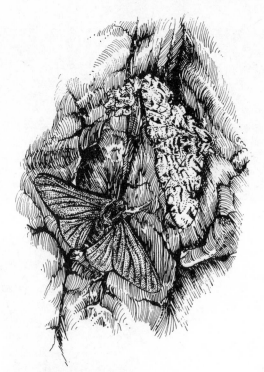

PEPPERED MOTH

This species originally came by the name 'peppered moth' because it had black spots on a white background. At one time the light coloured form was dominant, but with increasing pollution the rare black form increased in numbers and became more frequently encountered than its light-coloured counterpart.

two groups. The material on which the two groups feed determines, to some extent, their speed of movement. Millipedes are herbivores; centipedes are carnivores. The former move in an almost leisurely fashion, as they do not have actively to hunt their food. On the other hand, centipedes rely on a turn of speed to catch and kill live food.

Leaf litter animals

Millipedes will often be found in among the leaf litter which, in many instances, forms the basis of their diet. One common species is the snake millipede (*Cylindroiulus* spp.). At first sight it is not unreasonable for some people to confuse it with the also well-known wireworm. When not out looking for food, it may be found coiled up, rather like a watch spring. This species is brownish-black. Although it is equipped with considerably more legs than the centipede, it does not move any more quickly. As the legs move, they follow each other segment by segment, so that one gets the impression of a wave-like movement. The other millipedes likely to be found are those with black, almost shiny, bodies.

Millipedes are herbivores, and while centipedes have strong jaws, equipped with poison fangs, the jaws of the millipede, dealing with soft plant material, are not so well developed. Most millipedes feed on decaying material, providing a useful service in the garden, but those which do have a liking for root crops are hardly welcome. Attacks usually occur in dry weather, probably when the millipedes seek moisture which has ceased to be available from their usual sources, but once they have found a suitable supply they are hardly likely to leave it and so will do great damage. For most of the time, however, millipedes perform a useful service in the garden, particularly in refuse and compost heaps, where, by attacking the material, they bring about and increase the decaying process.

Victims to many predators

Because of their slow-moving nature, many millipedes will fall prey to other species. However, they do possess a means of repelling attackers. Their defence mechanism consists of stink glands, which produce a variety of evil-smelling liquids, containing hydrogen

BARN OWL
With its silent flight and ghostly appearance, the barn owl was generally feared by our ancestors, especially when it was seen in churchyards. At one time it found a plentiful supply of nesting sites in towns, laying its eggs in disused buildings and in hollow trees. Both of these nesting sites have decreased in numbers, and the birds has suffered as a result of this and from the use of pesticides.

cyanide! Although these liquids often do the job successfully they are not one hundred per cent effective. Various predators do not appear to be affected by the millipede's liquids. Indeed, not only do small mammals take them, but both toads and frogs are partial to the occasional millipede. Starlings feed on them, and spiders will also make their attacks.

The breeding season of the millipedes found in the British Isles will last for much of the summer. After mating, although all eggs are placed in the soil, their fate varies from one species to another. Some females will take the precaution of actually providing a nest for the eggs, most making this from their own droppings. But some species are not as meticulous: they drop the eggs in the soil, leaving them to develop. Some manage to survive, although many must be devoured.

Once the young millipedes are born, they will need to go through a series of moults before they take on the appearance of the adults which produced them. It is at the various moult stages that the young millipedes are most likely to fall prey to predatory species. Their soft bodies are easily attacked. Some take precautions by secreting themselves in nests; others leave themselves open to the many onslaughts which follow.

BLACK REDSTART (*Phoenicurus ochruros*)
The black redstart has a very special place in ornithological circles. It is the only rare British bird which, until recently, lived solely in built-up areas. Until fifty or so years ago, the black redstart has bred only a couple of times. A regular breeding pattern was established in 1923 when the bird bred in a south coast town. Although numbers rose to around 100 pairs as the bombed sites of London and other cities provided a place for it to search for food, there has again been a drastic decrease in its population.

The breeding population is now put at about twenty pairs, but it varies from year to year. Industrial areas, including power stations, seem to attract the bird, and it has actually bred inside Dungeness power station.

One of the black redstart's distinctive features is its almost startling chestnut coloured tail. It is from this that the bird gets the 'redstart' in its name. The Old English for a tail is *steart*.

In towns nests are placed in crevices, on ledges and under rafters, and are made from a mixture of roots, moss and grasses.

Large numbers of insects feature in the bird's diet, many of them being caught on the wing. It also takes other invertebrates, including millipedes, and spiders, as well as some berries, probably when other food is not easy to come by.

Carnivorous centipedes

Of the centipedes, the species most likely to be found is *Lithobius forficatus*. *Lithobius* means 'stone dweller', a reference to the place where the animal is often encountered. Because they could quickly dry up, centipedes come out at night, when they are not subject to the effects of the sun. Both centipedes and millipedes were placed in one group, the Myriapodia ('many feet'). Although the prefixes 'centi' and 'milli' suggest exceedingly large numbers of legs, this is false. On the body of millipedes, there are two pairs of legs to almost all of the body segments; on centipedes there is only one pair per segment.

Needing to kill live food, the centipede is equipped with claws to catch its prey. It then injects the unfortunate victim with a strong poison which kills the animal. Its diet consists of a wide variety of food, including mites, beetles, spiders, springtails, woodlice, slugs and worms. It appears that a few animals catch and eat centipedes, because the amount of poison they are able to produce is not enough to do any damage. Small mammals, some birds, toads and larger insects all take various centipedes from the smallest to the largest species.

As with some of our common animals, not a great deal is known about the life history of centipedes. As far as mating is concerned, it is thought that this takes place from mid-spring through the summer months. Because the many eggs are small and sticky they quickly disappear into the soil. Some species are better

BADGER

The badger—affectionately known as Brock— is learning to adapt itself to the urban life-style. The badger is becoming increasingly used to the noise and bustle of built-up areas. It is possible that it will change its feeding habits, often coming out later, since the town night-life ceases much later than that of the countryside.

parents than others, tending both their eggs and offspring. As with millipedes, once the young hatch they will only resemble the adults after several moults.

Favourite of the night

The badger (*Meles meles*) is a true nocturnal animal, and to most urban dwellers conjures up pictures of idyllic countryside, where the mammal will live away from habitations, mainly in wooded areas; but like so many other creatures it has changed its ways, and in many areas now copes with man in the town, provided of course that it can find a suitable area in which to live and breed. It is safe to assume that most will be found around the periphery of the urban environment. They need soil which can be excavated successfully, to form the underground home which occupies more space than that required by other mammals, including the fox, which may be found deep inside the heart of many towns.

However, once the badger has established itself, close to the town, or where the town encroaches on its home, it will soon take advantage of man. It will leave the confines of its sett to move into suitable urban environs, where it will enter gardens and even raid dustbins and rubbish tips. Although shy and secretive, it manages to adapt well and survive in areas where it is continually disturbed.

Young badgers tend to stay with the parents, excavating new homes within the existing sett if it is big enough. If not they will be expelled before the next litter is born. Where such a sett is in a confined area, and will not take further expansion, the youngsters will need to seek out new areas as excavation sites for a new sett. Since they have a preference for easily worked soils – sands and loams and chalks probably supporting the greatest number of setts – they will only establish new homes where these conditions occur.

Towns are not ideal

Of course the town and city do not provide the badger with ideal situations, but it is forced to adapt to the prevailing conditions, or die out. Thus, within the town, although setts are usually not as large as those in undisturbed woodland, the animal has found areas around canals, the vast expanses which rubbish tips offer, railway embankments and areas of waste ground. More recently, where urban motorways have developed, these are often accompanied by banked sites. This is a new habitat which the badger has exploited to its own advantage in some parts of the country.

Why a particular site is chosen is something of a mystery. As with other species, food is of paramount importance, and so it seems likely that the animal will investigate this aspect before moving into a new area.

In adapting to the urban situation, not only has the badger learned to live with man, but it has also changed its habits. In the country, the badger emerges at dusk, but here the area is likely to be undisturbed; this will not be the case in the town. Disturbances are likely to continue until later in the evening, and so the nightly foraging session seems to start much later. Like so many of the animals described in this book, the badger has learnt to adapt to live in the town and exist side by side with us. In many ways the ever-observant animals know a great deal more about our habits than most of us know about theirs!

Common ivy

Quite often we overlook many of our common plants. Sometimes we are ignorant of their value to wildlife. One plant which flourishes both on the ground and around older trees in churchyards is the ivy. It also manages to find a suitable support on walls, if it is allowed to grow there!

In spite of its great value to wildlife – it provides shelter and food – there are those people who believe that its disadvantages outweigh the advantages. In this case the plant is often destroyed. In the churchyard, however, it is usually much safer than in other areas. There is a view held by some foresters that ivy covering a tree trunk prevents the tree from growing properly. Furthermore the ivy roots are said to compete for water in the soil. In most conditions this is not likely to have a great deal of effect on the host tree.

Although ivy is evident throughout the year because it is evergreen, it is during the autumn and winter that the plant becomes especially conspicuous. With the leaves gone from its host, the glossy green foliage of the ivy stands out. An examination of its leaves will soon give some indication of how it manages to survive in what, for many other species, are particularly unfavourable conditions. The wax-like covering enables the ivy to retain moisture during adverse spells. In exceptionally severe winters the ivy is likely to suffer from 'frost-bite'. Some of its flowering shoots will die.

One of the interesting aspects of the ivy is the variety of leaf shape. Basically there are leaves which have either three or four points or lobes. An examination of the leaves which appear on the flowering shoots will reveal that they have an outline which is generally egg-shaped. It is on the growing shoots that the lobed leaves can be seen.

Generally dark green in colour, the leaves have a waxy, almost leathery, feel to them. The upper surface exhibits a dark green sheen; the under-surface is lighter. There is an attractive pattern to every leaf caused by the lighter colour of the veins.

Like other green plants ivy needs sunlight to make its

food. Where the plant grows on a flat surface such as a church wall, there is little overlap of the leaves. This ensures maximum exposure to the sun. This is not always possible, though, where ivy grows on trees, and here a great deal of overlapping occurs.

In churchyards birds may drop the seeds which will germinate and although relatively slow-growing in its early stages the ivy soon becomes established. It will grow almost anywhere where it can gain a root-hold, but flowering and fruiting will only take place where the plant is in sunlight. In deep shade it grows well, but never produces fruits.

Not all ivy manages to climb. In some neglected corner of a churchyard the plant may sprawl over the ground. Here it might appear less vigorous than when it can find support for upward growth. Nevertheless it can rapidly colonise bare ground.

It is in the autumn and winter that ivy is especially valuable to wildlife. When most other species have lost their flowers the ivy is just beginning to produce its supply. Although the flower buds will probably be on the shrub from about the middle of summer they are often overlooked. The earliest flowers can probably be seen sometime during September, with a great profusion appearing in October and November. Some can still be seen until the end of the year.

Ivy is a magnet for autumn insects. A rich and seemingly endless supply of nectar exudes from the yellow-green flowers. Wasps, virtually at the end of their life, still eagerly seek out nourishment and come to the ivy. Bees and butterflies, along with many other marauding insects, will seek out a meal. On mild autumn nights moths can be seen hovering around the ivy flowers, ever-eager for a meal of the sweet tasting liquid.

The insects are important to the ivy because they will bring about pollination. In due course the berries will appear. They ripen slowly, and by next spring will offer tasty morsels for such species as robins, blackbirds and mistle thrushes.

LONG-EARED BAT
The long-eared bat will leave its roost about half an hour before sunset. It will probably feed for a while, return to rest, and then re-emerge to take further exercise and food. It is possible that it will do this throughout the night, until either its appetite is satisfied or dawn arrives.

The dense cover which ivy provides is eagerly sought out by some hibernating animals. Small tortoiseshell butterflies will seek shelter. Birds find a refuge, as do some small mammals.

Lichens
Like the yew, the churchyard also harbours another long-lived plant, albeit much less spectacular. It is home to a variety of species of lichens. Lichens are strange, unique plants, because each is made up of two different plants – a fungus and an alga. In the churchyards the stonework of both graves and of the church itself provides a home for these slow-growing plants.

Until recently most lichens were only known by their Latin names. Many now have a common name, however. Grey warted lichen (*Physica caesia*) is found on tombstones and is one species which is not too badly affected by pollution, a problem which eradicates some species. The grey warted lichen grows best on stones where bird droppings accumulate, especially in sites where there is also plenty of light.

The mustard lichen (*Candalariella vitellina*) is, as its common name implies, mustard-coloured. Small patches occur on walls as well as on rock. Swedish people collected the lichen, extracted a dye, and then used this to colour their church candles.

Some species, like the frosted wall lichen (*Buellia canascens*) may find a variety of places to grow from tree trunks to walls. Although it is common in southern parts of Britain, it is much less frequently encountered forther north, becoming rare in north Scotland.

Another species frequently found on tombstones is the orange star lichen (*Coloplaca heppiana*). This particular species grows in such a way that it forms a rosette on the gravestone.

The letters on gravestones are often picked out as the sulphur lichen develops. Once established on a tombstone, it may gradually cover the whole of the surface. It seems to grow better where there is moisture which often collects in the engravings.

Longhorn beetles

The longhorn is a group of beetles which may live in churchyards. Requiring wood, there is usually a supply in the churchyard. Once the female is ready to lay her eggs she will search out suitably deep cracks in the bark of trees. Once the eggs have been deposited, the female longhorn will often produce a quick-drying viscous liquid. This effectively seals in the eggs, and helps to protect them.

The eggs will remain in the wood for two or three weeks, and will then hatch. They are cylindrical in outline and they have a white skin. Wood will provide them with nourishment, and to eat the wood the longhorn larvae are equipped with strong jaws. Both living and dead wood provides a source of food, and where the larvae have been there will be a wide tunnel.

Having spent two or three years deep inside the wood, the larvae will be ready to pupate, and will make their way towards the surface. Pupation will take place in a cell close to the exterior. Although the average period may be two or three years, this is by no means normal. Wood which has been cut down, seasoned, and then made into furniture has harboured longhorn beetles larvae. In some cases it was twenty years after the furniture was made that the adult beetles emerged!

The adults need to take nourishment, and it is likely that many feed on nectar. Various species of longhorns – and there are sixty-five in Britain – may land on umbelliers, where they imbibe the nutritious nectar. Some species eat the leaves of the tree in which the female lays her eggs.

One species of longhorn beetle, called the wasp beetle because of its yellow and black markings, is quite common. Although only about 14mm ($\frac{5}{8}$in) in length its bright colour makes it easily recognisable. It is found out and about in churchyards, as well as in gardens and parks in May and June. A variety of wood will provide the larvae with food, and once the eggs have been laid the emerging larvae feed for up to two years.

Leaf-litter dwellers

Nature is so adaptable and so versatile that plants and animals survive almost everywhere. Some of this wildlife is obvious to the human eye, but in the churchyard, as elsewhere, much goes on year in and year out undetected by the passer-by.

Such is the case of leaf litter. The debris which accumulates under trees, especially in odd and unforgotten corners of the churchyard, provides home and food for many creatures. Those species which make their homes permanently in the litter will in turn provide a supply of other creatures with food, should they be passing by. Numbers of hunting and wolf spiders, belonging to the Lycosidae family, are frequently encountered in the litter. Brown in colour, they are well camouflaged. Because they do not make webs in which to trap their food, they have to hunt actively for their prey. Speed is of the essence if the spiders are to catch creatures living in the litter. Where there is a good collection of dried leaves it is possible to hear them scuttling about on a dry, summer day.

It is important that creatures live in the leaf litter since it has to be broken down. The leaves and other debris must decay so that the nutrients once removed from the soil will be returned, allowing the cycle of events to continue. A large number of mites, of a variety of species, live here. Some are large enough to be seen by the human eye. Others are much smaller. The mites help to begin the process of decay, and are assisted by other leaf-litter dwellers and visitors.

A search through an accumulated supply of leaf litter will usually reveal a variety of insects. The majority of these will be visiting, passing through, as it were. There are few insects which are exclusively confined to the leaf litter. Several species – all lacking the power of flight, which is a characteristic of many other insect species – live here in some form of security. Among these primitive insects are the springtails (*see* page 25), two-pronged bristletails, three-pronged bristletails and proturans.

The last three groups are all very small. Two-pronged bristletails are white, having long antennae but no eyes. Three-pronged bristletails also have long antennae, but with compound eyes they are able to see.

You might discover ladybirds in leaf litter, especially in the autumn. It appears that the layer of leaves provides them with a measure of protection at this time of the year. Some species of two groups of beetles – rove beetles and ground beetles – will live on the leaf litter all the while. Some are carnivores, killing other animals which live there. Others are herbivores, feeding on the dead and decaying material.

INDEX

Figures in **bold** refer to page numbers of illustrations.